Springer Tracts in Mechanical Engineering

About this Series

Springer Tracts in Mechanical Engineering (STME) publishes the latest developments in Mechanical Engineering - quickly, informally and with high quality. The intent is to cover all the main branches of mechanical engineering, both theoretical and applied, including:

- Engineering Design
- Machinery and Machine Elements
- Mechanical structures and stress analysis
- Automotive Engineering
- Engine Technology
- Aerospace Technology and Astronautics
- Nanotechnology and Microengineering
- Control, Robotics, Mechatronics
- MEMS
- Theoretical and Applied Mechanics
- Dynamical Systems, Control
- Fluids mechanics
- Engineering Thermodynamics, Heat and Mass Transfer
- Manufacturing
- Precision engineering, Instrumentation, Measurement
- Materials Engineering
- Tribology and surface technology

Within the scopes of the series are monographs, professional books or graduate textbooks, edited volumes as well as outstanding PhD theses and books purposely devoted to support education in mechanical engineering at graduate and post-graduate levels.

More information about this series at
http://www.springer.com/series/11693

Esteban Ferrer • Adeline Montlaur
Editors

CFD for Wind and Tidal Offshore Turbines

Springer

Editors
Esteban Ferrer
School of Aeronautics
Universidad Politecnica de Madrid (ETSIA-UPM)
Madrid
Spain

Adeline Montlaur
Universitat Politecnica de Catalunya
Escola d'Enginyeria de Telecomunicacio i Aeronautica
Catelldefels
Spain

ISSN 2195-9862 ISSN 2195-9870 (electronic)
Springer Tracts in Mechanical Engineering
ISBN 978-3-319-36561-9 ISBN 978-3-319-16202-7 (eBook)
DOI 10.1007/978-3-319-16202-7

Springer Cham Heidelberg New York Dordrecht London

Softcover reprint of the hardcover 1st edition 2015

Printed on acid-free paper

Springer International Publishing AG Switzerland is part of Springer Science+Business Media (www.springer.com)

Introduction

The International Energy Agency (IEA) concludes in The World Energy Outlook 2008 that the current energy consumption and production is "patently unsustainable environmentally, economically, and socially". Social concern and international agreements (e.g. Kyoto 1997, Copenhagen 2009 or Durban 2011) are pushing forward the development of renewable energy technologies for sustainable and renewable energy generation.

In particular, offshore wind and tidal turbines have seen increasing interest from academia, industry and government bodies, during recent years, as offshore sites present huge energy potential. The new engineering challenges presented by these technologies, together with the difficulty to undertake experimental test under offshore environments, have raised the interest on Computational Fluid Dynamics (CFD) to design appropriate turbines and blades, understand fluid flow physical phenomena associated with offshore environments and predict power production, among others.

This book encompasses novel CFD techniques to compute offshore wind and tidal applications. All the included papers have been presented at the 11th World Congress on Computational Mechanics (WCCM XI) organised together with the 6th European Conference on Computational Fluid Dynamics (ECFD VI) in Barcelona 2014. The book includes contributions of researchers from academia and industry.

December 2014
Barcelona, Spain

Esteban Ferrer
Adeline de Montlaur

Contents

Chapter 1
Flow Scales in Cross-Flow Turbines

Esteban Ferrer and Soledad Le Clainche

Abstract This work presents analytical estimates for various flow scales encountered in cross-flow turbines (i.e. Darrieus type or vertical axis) for renewable energy generation (both wind and tidal). These estimates enable the exploration of spatial or temporal interactions between flow phenomena and provide quantitative and qualitative bounds of the three main flow phenomena: the foil scale, the vortex scale and wake scale. Finally using the scale analysis, we show using an illustrative example how high order computational methods prove beneficial when solving the flow physics involved in cross-flow turbines.

1.1 Introduction

Problems where the forces on rotating or oscillating bodies in a fluid are to be predicted are common in engineering applications and result in fluid–structure interaction situations. Examples are flows around isolated rotating bodies and foils, turbomachinery applications, insect flight aerodynamics, unmanned air vehicles and, more recently, flows through renewable energy devices, e.g. wind and tidal turbines.

A particularly challenging problem is presented by cross-flow wind and tidal turbines for power generation (also known as vertical-axis, H-rotors or Darrieus turbines). These types of turbine consist of foil shaped blades that generate lift forces so as to rotate a shaft to which the blades are connected. Therefore azimuthal changes in blade aerodynamics (or hydrodynamics)[1] are common, resulting in

[1]Airfoil aerodynamics and hydrofoil hydrodynamics are equivalent nomenclatures for foils operating in air or water environments. Since this work encompasses both wind and tidal turbine applications, from this point onwards, "foils" will denote either "airfoils" or "hydrofoils". In addition, the term "aerodynamic" can always be replaced by "hydrodynamic" in this work.

E. Ferrer (✉) • S. Le Clainche
ETSIA-UPM - School of Aeronautics, Plaza Cardenal Cisneros 3, 28040 Madrid, Spain
e-mail: esteban.ferrer@upm.es; soledad.leclainche@upm.es

E. Ferrer, A. Montlaur (eds.), *CFD for Wind and Tidal Offshore Turbines*,
Springer Tracts in Mechanical Engineering, DOI 10.1007/978-3-319-16202-7_1

complex flow phenomena such as stalled flows, vortex shedding and blade–vortex interactions.

Cross-Flow Turbines (CFT) are sometimes referred to as vertical-axis turbines, however the term cross-flow turbine is preferred since the absolute turbine position is omitted and the relative flow-axis geometry is emphasised through this terminology.

To date, this type of turbine configuration has had limited use within the wind energy sector, where the three bladed axial flow turbine has been widely adopted. However, it is thought that CFT configurations can be advantageous for new emergent markets as offshore wind and tidal energy production and also for installation in urban environments. A brief review of some of the arguments in favour and against this type of configuration follows. A more complete discussion can be found in [6].

On the one hand, the main drawback is that CFT are generally less efficient than axial flow turbines since the downstream half of the turbine produces less torque due to the shadowing from the upstream blades. Furthermore, since for part of the cycle the blade moves parallel to the flow, the lift force powering the blades, being proportional to the incident flow speed squared, is reduced.

On the other hand, this type of device has the advantage of not requiring a specific orientation relative to the flow, since it can work independently of the stream direction and hence some researchers argue that this technology is more suitable for the offshore environment (offshore wind and tidal) as it would minimise maintenance costs through reduced control systems (e.g. yaw mechanism) [11, 12]. Furthermore, due to its geometrical simplicity, the generator can be located far from the rotating blades, reducing installation complexity and maintenance cost.

To the authors' knowledge, the analysis presented herein is a first attempt to derive analytical estimates that may be used in the future in simplified models of blade element momentum type. The analytical bounds included in this study may provide limiting conditions for analytical models (e.g. [16, 20, 21, 25] for cross-flow wind turbines or [5, 13, 24] in the tidal energy context).

This chapter characterises various physical phenomena encountered in cross-flow turbines and quantifies its spatial and temporal importance in terms of length, time and velocity scales. The quantification of the various physical phenomena provides guidelines for the appropriate simulation of CFT using numerical techniques (e.g. spatial and temporal resolution). Finally, in Sect. 1.4, we address the necessity of high accuracy when computing cross-flow turbine flows using numerical techniques, such as the high order Discontinuous Galerkin method developed by the first author [7–10].

1.2 Physical Characterisation of Cross-Flow Turbine Flows

Both wind and tidal CFT technologies operate at relatively low flow speed, when compared to the speed of sound in their respective media, minimising the effects of fluid compressibility. Typical CFT flows can be characterised using the incompressible Navier–Stokes equations. This system of non-linear partial differential equations can be written in its convective form as the set:

$$\text{Continuity}: \qquad \nabla \cdot \mathbf{u} = 0, \tag{1.1}$$

$$\text{Momentum}: \qquad \frac{\partial \mathbf{u}}{\partial t} + (\mathbf{u} \cdot \nabla)\mathbf{u} = -\frac{1}{\varrho}\nabla p + \nu \nabla^2 \mathbf{u}, \tag{1.2}$$

where $\mathbf{u} = (u, v, w)^T$ is the vector of the velocity components in three dimensions, p represents the static pressure, ϱ is the density and $\nu = \frac{\mu}{\varrho}$ is the kinematic viscosity, with μ the dynamic viscosity.

The previously defined Navier–Stokes equations can be written in non-dimensional form using the following non-dimensional variables:

$$\mathbf{u}^* = \frac{\mathbf{u}}{U}, \quad p^* = \frac{p}{\varrho U^2}, \quad t^* = t\frac{U}{L}, \quad \nabla^* = L\nabla, \tag{1.3}$$

where, in addition to the defined variables, the following scalars are introduced: U as the characteristic flow velocity (e.g. the freestream speed) and L as the characteristic flow length. Substitution of the non-dimensional quantities into Eq. (1.2) leads to a non-dimensional momentum equation:

$$\frac{\partial \mathbf{u}^*}{\partial t^*} + (\mathbf{u}^* \cdot \nabla^*)\mathbf{u}^* = -\nabla^* p^* + \frac{1}{\text{Re}}\nabla^{*2}\mathbf{u}^*, \tag{1.4}$$

where $\text{Re} = LU/\nu$ is the Reynolds number.

Let us estimate the Reynolds numbers to characterise the flows for wind and tidal turbines. Table 1.1 shows characteristic fluid properties for air (at 15 °C) and sea water (at 0 °C and moderate salinity). The table also includes typical mean flow speeds at wind [3] and tidal sites [5, 18]. As can be seen, both fluids are similar in

Table 1.1 Air and sea water fluid and flow properties

	Units	Air	Sea water
Fluid property			
Density (ϱ)	kg/m^3	~1.225	~1,027
Dynamic viscosity (μ)	Pa s	~1.78 × 10^{-5}	~1.88 × 10^{-3}
Kinematic viscosity (ν)	m^2/s	~1.45 × 10^{-5}	~1.83 × 10^{-6}
Flow property			
Mean flow speed	m/s	~12	~1.5

terms of their kinematics: $U_{\text{air}}/\nu_{\text{air}} \sim U_{\text{sea}}/\nu_{\text{sea}}$. To estimate the Reynolds numbers, let us consider a small cross-flow turbine of radius $R = 5\,\text{m}$ (e.g. a wind turbine for urban environment and a typical tidal turbine), then the Reynolds number Re_D based on the turbine diameter D results in ${\text{Re}_D}^{\text{air}} \approx {\text{Re}_D}^{\text{water}} \approx 4 \times 10^6$, showing that from a flow dynamic point of view both technologies operate in similar regimes.

Remark When tidal turbine flows are considered, the Froude number $\text{Fr} = U/\sqrt{gL}$ may be of importance. However for mildly blocked configurations (i.e. low ratio of turbine diameter to shallow water depth) this parameter can be shown to have a relatively low importance [5, 15] and will not be considered in this chapter.

1.3 Disparity of Flow Scales and Their Simulation

This section describes the range of flow scales expected when simulating cross-flow turbine flows in relation to the resolution accuracy required for its simulation. Figure 1.1 presents a schematic illustration of the three physical spatial scales expected on CFT flows: the foil scale, the vortex scale and wake scale. The foil scaling includes near wall scales and boundary layers, the vortex scaling considers vortex shedding and blade–vortex interaction effects and finally the wake scaling includes shadowing effects due to the turbine blockage [5, 15, 23] and the wake structure related to the rotating blades. In this section, estimates for the length, time and velocity scales for the different phenomena are introduced.

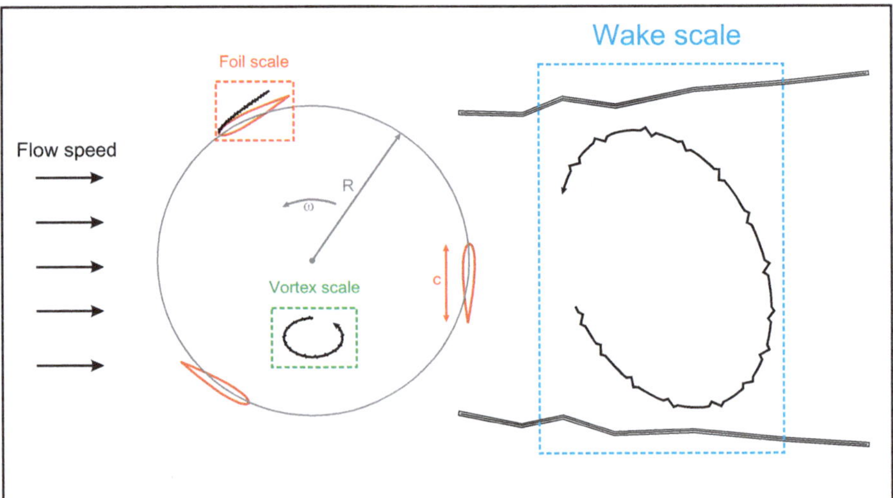

Fig. 1.1 Flow scales in cross-flow turbine flows. Turbine blades of chord c rotate with an angular velocity ω at a distance R from the turbine centre

1.3.1 The Foil Scaling

In the presence of a non-slip condition at the wall surface, the flow is forced to slow down to have zero velocity at the wall. This creates the so-called boundary layer (defined by Prandtl in 1904) that will develop on turbine blades. The logarithmic profile of tangential velocities developed at the wall (if attached flow is considered) has to be properly modelled in a numerical simulation, since it directly influences the aerodynamic forces on the blades. Typically, the foil boundary layer has a laminar portion which will undergo transition to turbulence some distance downstream (assuming a high enough Reynolds number); this distance being influenced by, among other factors, the pressure gradient on the surface, the wall surface roughness and the free stream turbulence outside the boundary layer. Turbulent boundary layers are thicker and carry more momentum and thus separate later than laminar boundary layers. Boundary layers are a vast topic of research and only the principal characteristics have been outlined here. For a detailed explanation the reader is referred to the monograph by Schlichting [19].

It is possible to estimate CFT's length scales for the boundary layer using the Blasius approximation [19] for the flow over a flat plate with zero pressure gradient. First, let us consider the rotational speed of the blades $U_{\text{rot}} = \omega R$, where ω is the angular velocity and R the turbine radius. In addition, we define the tip speed ratio as $\lambda = U_{\text{rot}}/U$, where U represents the free stream velocity. For simplicity, the velocity that the foil experiences can be estimated as $U_{\text{foil}} = \sqrt{U_{\text{rot}}^2 + U^2} \approx U_{\text{rot}} = \lambda U$, which is a valid assumption for $\lambda >> 1$ (i.e. ignoring the free stream velocity U). Then, Blasius' approximation provides a boundary layer of thickness: $\delta \approx x/(\frac{xU_{\text{foil}}}{\nu})^{\gamma} = x/(\frac{x\lambda U}{\nu})^{\gamma}$, where x is the distance from the leading edge and the exponent γ relates to the boundary layer regime: $\gamma = 1/2$ for a laminar regime and $\gamma = 1/5$ for a turbulent boundary layer. By setting $x = c$ (i.e. the foil chord), it is possible to obtain a characteristic length scale for the foil $\ell_{\text{foil}} = \frac{c}{(\text{Re}_{\text{foil}})^{\gamma}}$, where $\text{Re}_{\text{foil}} = \frac{cU_{\text{foil}}}{\nu}$ is the Reynolds number based on the foil chord. An estimate for the time scale can be readily found: $\tau_{\text{foil}} = \frac{\ell_{\text{foil}}}{U_{\text{foil}}} = \frac{c}{U\lambda(\text{Re}_{\text{foil}})^{\gamma}}$. A summary of these scales is given in Table 1.2.

Table 1.2 Characteristic foil scales in CFT turbines

Length scale ℓ_{foil}	Time scale τ_{foil}	Velocity scale U_{foil}	Reynolds number Re_{foil}
$\frac{c}{(\text{Re}_{\text{foil}})^{\gamma}}$	$\frac{c}{U\lambda(\text{Re}_{\text{foil}})^{\gamma}}$	λU	$\frac{c\lambda U}{\nu}$

1.3.2 The Vortex Scaling and Blade–Vortex Interaction

Vortex shedding and vortex impingement may play an important role in turbine performance. For cross-flow turbines, a vortex created by the upstream half of the turbine may be convected downstream, interacting with the rear half of the turbine. Numerical accuracy can significantly influence this phenomenon since vortices may diffuse and lose their structure, changing the forces seen by the rear passing blade. Typically, low order methods (e.g. Finite Volume) have been used to compute CFT flows (e.g. [4, 14, 17]), however, these show non-negligible diffusive and dispersive errors that may disturb flow structures (e.g. vortices or wakes, see [7] for illustrations). In numerical simulations, high order methods (with low numerical errors, e.g. [7, 9]) codes are preferred to capture accurately vortex evolution and blade–vortex interaction phenomena.

An estimate for the scales corresponding to the vortex convection and the blade–vortex interaction phenomenon is provided. Firstly, let us assume a two dimensional vortex with area $A \approx \frac{\pi}{4}c^2$ (the circular area covered by the two dimensional vortex), where c is the characteristic length (the foil chord and the diameter of the vortex) and with characteristic turn over time $\tau_{\text{vortex}} \approx \frac{2\pi}{\omega_z}$, where ω_z represents the vorticity magnitude. By approximating the vorticity as a function of the foil circulation and thus of the lift generated by the foil (Kutta–Joukowski theorem [2]), it is possible to write $\omega_z \approx \frac{\Gamma}{A}$, with $\Gamma = \frac{L}{\varrho U_{\text{foil}}}$ representing the circulation and L the foil lift force. Since $L = \frac{1}{2}\varrho U_{\text{foil}}^2 c C_L$, with C_L the foil lift coefficient, the characteristic vortex time can be expressed as: $\tau_{\text{vortex}} \approx \frac{\pi^2 c}{\lambda U C_L}$. Finally the characteristic velocity of the vortex can be estimated as $U_{\text{vortex}} = \frac{\ell_{\text{vortex}}}{\tau_{\text{vortex}}} \approx \frac{\lambda U C_L}{\pi^2}$.

To estimate the lengths and times of the blade–vortex (BV) interaction phenomenon, let us assume a vortex of diameter c that convects through the turbine at a velocity $U_{\text{axial}} = \beta U$, where $\beta < 1$ represents the velocity deficit through the turbine[2] and U_{axial} is the mean streamwise velocity component within the turbine. The time for a vortex to convect from the front half to the rear half (i.e. to travel the distance $D = 2R$) can be estimated as $\tau_{BV} = \frac{\ell_{BV}}{U_{\text{axial}}} = \frac{D}{\beta U}$. Table 1.3 summarises these scales.

Table 1.3 Characteristic vortex scales in CFT turbines

Length scale	Time scale	Velocity scale	Reynolds number
ℓ_{vortex}	τ_{vortex}	U_{vortex}	$\text{Re}_{\text{vortex}}$
c	$\frac{\pi^2 c}{(\lambda U C_L)}$	$\frac{\lambda U C_L}{\pi^2}$	$\frac{c\lambda U C_L}{\pi^2 \nu}$
ℓ_{BV}	τ_{BV}	U_{BV}	Re_{BV}
D	$\frac{D}{\beta U}$	βU	$\frac{D\beta U}{\nu}$

[2] $\beta < 1$ for unblocked cases but may exceed 1 if blockage leads to accelerated flow through the turbine.

Table 1.4 Characteristic wake scales in CFT turbines

Length scale ℓ_{wake}	Time scale τ_{wake}	Velocity scale U_{wake}	Reynolds number Re_{wake}
D	$\frac{D}{\beta\beta' U}$	$\beta\beta' U$	$\frac{D\beta\beta' U}{\nu}$

1.3.3 The Wake Scaling

Cross-flow turbine wake development and dissipation are of major importance since the performance of downstream turbines may be influenced by the structure and strength of upstream generated structures. Furthermore, the wake development modifies the incoming flow speed and consequently angle of attack of the turbine blades. This may alter the performance and efficiency of the turbine and thus its correct prediction is important. Unconfined wakes have been studied for a long time in the context of wind turbines. For example, the paper by Vermeer et al. [22] is an extensive review of experimental and numerical results for axial wind turbine wakes. Although various empirical models exist to predict the extension of the wake behind an operating wind turbine, and many numerical codes have been used to describe the wake structure, the study concludes that more experimental data is needed and that the wake characteristics are still not well understood. However, it seems clear that the wake can be subdivided into a near wake where coherent structures develop, and a far wake, where mixing generates an almost uniform velocity deficit.

Let us estimate the length and time scales for the near wake assuming a wake of diameter $D = 2R$, and a velocity in the wake region as $U_{\text{wake}} = \beta\beta' U$, where in addition to the induction factor β introduced in the previous section, $\beta' < 1$ is defined as the velocity deficit due to the rear half blade passage. The characteristic scales are detailed in Table 1.4.

1.3.4 Scale Comparison and Numerical Resolution

It is possible to compare the different length and time scales obtained for the different phenomena (Tables 1.2, 1.3 and 1.4) to estimate the zones that require more spatial and temporal resolution for a numerical simulation. The most stringent length and time scales are clearly the foil scale since both decrease as the Reynolds number $\mathrm{Re}_{\text{foil}}$ increases. The vortex time scale is the smallest after the foil time scale but of the same order of magnitude as the blade–vortex interaction scale.

When comparing the time scales for the blade–vortex interaction and the near wake resolution, one can see that $\ell_{\text{wake}} \approx \ell_{BV}$ and $\tau_{\text{wake}} \approx \frac{\tau_{BV}}{\beta'}$, which implies that for $\beta' < 1$ the blade–vortex interaction phenomenon requires smaller temporal resolution than the wake resolution.

In addition, one may examine the factors influencing the separation of scales using the ratios of the smallest to largest scales. Let us define the ratio of length scales as $r_\ell = \frac{\ell_{\text{foil}}}{\ell_{\text{wake}}}$. Substitution of the appropriate length scales (from Tables 1.2 and 1.4) lead to $r_\ell = \frac{c}{D(\text{Re}_{\text{foil}})^\gamma}$. This expression can be further simplified by using the definition of the turbine solidity $\sigma/N = 2c/D$, where N is the number of turbine blades. The final expression for the ratio of length scales reads: $r_\ell = \frac{\sigma}{2N(\text{Re}_{\text{foil}})^\gamma}$. Examination of the last expression shows that the separation of these length scales becomes larger (i.e. smaller ratio) if the number of blades (N) and the Reynolds number (Re_{foil}) increase or the solidity (σ) decreases.

Similarly, a ratio for the time scales can be derived leading to $r_\tau = \frac{\tau_{\text{foil}}}{\tau_{\text{wake}}} = \frac{\sigma\beta\beta'}{2N\lambda(\text{Re}_{\text{foil}})^\gamma}$. In this case, the time scale separation becomes larger if the number of blades (N), the Reynolds number (Re_{foil}) and the tip speed ratio (λ) increase, or the solidity (σ) and induction factors ($\beta\beta'$) decrease. It may be noted, from the derived expressions, that the time scale separation is more restrictive than the length scale separation since: $r_\tau = \frac{\beta\beta'}{\lambda} r_\ell$ with $\beta\beta' < 1$ and $\lambda >> 1$.

To conclude this section, a quantitative estimation of the various scales (Tables 1.2, 1.3 and 1.4) is provided in Table 1.5 for a typical CFT using the air and water environmental conditions introduced in Sect. 1.2. Namely, let us consider a cross-flow turbine of radius $R = 5$ m consisting of airfoils of chord $c = 1$ m (i.e. solidity $\sigma/N = 0.2$). To evaluate the length, time and velocity scales, it is necessary to approximate the following parameters: $\gamma = 1/5$ for a turbulent boundary layer, the tip speed ratio $\lambda \approx 3$, the lift coefficient $C_L \approx 1$ (e.g. maximum C_L for a NACA0015 at $\text{Re} = 1 \times 10^6$ [1]) and the induction factors $\beta \approx \beta' \approx 0.5$. In addition, calculations are performed for two flow speeds $U = 12$ and 1.5 m/s corresponding to air and water environments respectively (i.e. wind and tidal turbines). Note that the two media require different values for the kinematic viscosity: $\nu = 1.45 \times 10^{-5}$ and 1.83×10^{-6} m^2/s for air and salted water, respectively. Inspection of Table 1.5 reveals that smaller time scales are present

Table 1.5 Summary and evaluation of CFT scales for wind and tidal energy

	Length scale ℓ (m)	Time scale τ (s)	Velocity scale U (m/s)	Reynolds number Re
Wind turbine				
Foil	0.053	0.001	36.000	2.5×10^6
Vortex	1.000	0.274	3.648	2.5×10^5
BV	10.000	1.667	6.000	4.1×10^6
Wake	10.000	3.333	3.000	2.1×10^6
Tidal turbine				
Foil	0.053	0.012	4.500	2.5×10^6
Vortex	1.000	2.193	0.456	2.5×10^5
BV	10.000	13.333	0.750	4.1×10^6
Wake	10.000	26.667	0.375	2.0×10^6

when considering wind turbines over tidal turbines. However, it can be noted that the ratios of time scales (i.e. r_τ) are of the same order for both types of turbines.

The results from this simplified analysis confirm the intuitive reasoning on scale separation. However, it appears useful as it outlines the wide range of length and time scales that need to be accurately simulated to capture correctly the complex flow physics involved in cross-flow turbine simulations. In addition, the analysis allows for quantification of the relative importance of the various flow phenomena.

1.4 Engineering Accuracy and High Accuracy

It is often argued that high accuracy is not compulsory for most engineering applications since an error of 5–10 % may be acceptable and high order techniques are not necessary. In this section we show that the propagation of errors can result in unacceptable inaccuracies when evaluating the performance of CFTs. To exemplify why high accuracy may be of interest, the configuration depicted in Fig. 1.2 is considered. The figure depicts an array of two turbines where the second operates in the wake of the first.

Let us assume that the flow field around the first turbine is accurately resolved up to an error of 5 %: $U = U^{\text{exact}}(1 \pm 0.05)$. Assuming also a 5 % error when calculating U_{axial}, the error in the induction factor $\frac{U_{\text{axial}}}{U} = \beta$ (see Sect. 1.3) can be bounded as $\beta \approx \beta^{\text{exact}}(1 \pm 0.07)$. If the induction factor for the rear half $\beta' = \frac{U_{\text{wake}}}{U_{\text{axial}}}$ has the same errors associated than β (which is clearly a conservative estimate), one obtains $\beta' \approx \beta'^{\text{exact}}(1 \pm 0.07)$. The wake velocity can now be estimated following: $U_{\text{wake}} = \beta\beta' U$ and has bounds $U_{\text{wake}} = \beta^{\text{exact}}\beta'^{\text{exact}}U^{\text{exact}}(1 \pm 0.07)(1 \pm 0.07)(1 \pm 0.05) \approx U_{\text{wake}}^{\text{exact}}(1 \pm 0.11)$. This corresponds to a 11 % error in the incoming velocity for the second turbine. The examples show how an initially "engineering acceptable" error of 5 % may result in unacceptable levels after only one row of turbines. If a similar analysis is performed for the second row, it can be shown that the error increases to 15 % for the wake (assuming the same errors in the induction factors).

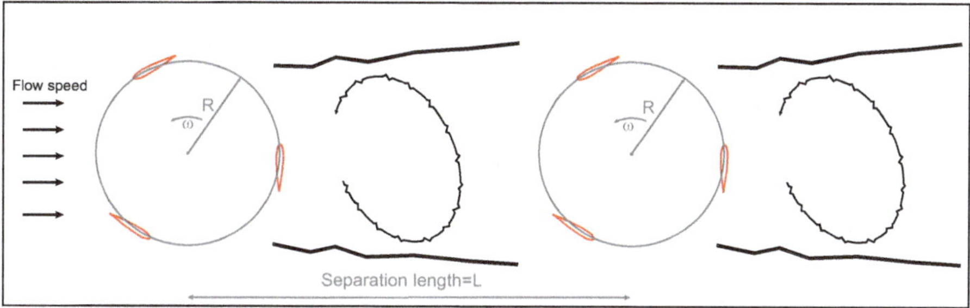

Fig. 1.2 Array of turbines (schematic)

Error bounds can also be estimated for blade forces and power output for these turbines. Let us consider an error when calculating the lift force of 7 %: $L = L^{\text{exact}}(1 \pm 0.07)$, which relates quadratically to the 5 % error in the velocity U. The power generated by the device is Power $= \omega T$, where T is the torque (i.e. the foil force in the tangential direction to the circular path) and ω is assumed to be an error-free rotational speed. For simplicity, let us state that $T = f(L)$, where f is a linear function depending mainly on the angle of attack. One can thus express the power as: Power $= \omega f(L^{\text{exact}}(1 \pm 0.07))$. It is common practice, in the turbine community, to calculate the non-dimensional coefficient of power $\text{Cp} = \frac{\text{Power}}{1/2\varrho A U^3}$, where $A = D = 2R$ represents the frontal area for a CFT, which results in $\text{Cp} = \text{Cp}^{\text{exact}} \frac{(1\pm0.07)}{(1\pm0.05)^3} \approx \text{Cp}^{\text{exact}}(1 \pm 0.11)$. Once more a 5 % error evolves into 11 % error in the variable of interest. As for the downstream turbine that sees, as its incoming flow, the wake velocity of the first turbine, the error in the power coefficient increases to 25 % $\left(\text{i.e. Cp} = \text{Cp}^{\text{exact}} \frac{(1\pm0.15)}{(1\pm0.11)^3}\right)$.

These two simple examples illustrate why high accuracy may be of interest. Because the physical phenomena and interactions (e.g. stalled flows, blade–vortex interaction) are much more involved than presented by this simplified approach, unacceptable error levels can be expected if low accuracy is used to compute complex physics related to cross-flow turbines.

1.5 Conclusions

This chapter details and characterises CFT flows in terms of time, length and velocity scales for the different phenomena encountered when simulating this type of devices. Namely, the foil, vortex and wake scales are characterised. Furthermore, the main factors influencing the separation of scales are analysed and used to weight the relative importance of various physical phenomena.

In addition, the necessity of high accuracy when computing CFT flows is addressed illustrating the beneficial potential of using numerical methods that provide accurate solution with minimum dissipative and dispersive errors (e.g. Spectral or Discontinuous Galerkin methods) such as the high order Discontinuous Galerkin code developed by the first author [7–10]).

Acknowledgements EF would like to acknowledge the financial support of the John Fell OUP fund and the European Commission (ANADE project under grant contract PITN-GA-289428).

References

1. Abbott I, Von-Doenhoff A (1959) Theory of wing sections: including a summary of airfoil data. Dover, New York
2. Anderson JD (2005) Fundamentals of aerodynamics, 4th edn. McGraw-Hill, New York

3. Burton T, Jenkins N, Sharpe D, Bossanyi E (2001) Wind energy handbook. Wiley, New York
4. Consul CA, Willden RHJ, Ferrer E, McCulloch MD (2009) Influence of solidity on the performance of a cross-flow turbine. In: Proceedings of the 8th European wave and tidal energy conference
5. Draper S (2011) Tidal stream energy extraction in coastal basins. Ph.D. thesis, University of Oxford
6. Eriksson S, Bernhoff H, Leijon M (2008) Evaluation of different turbine concepts for wind power. Renew Sust Energ Rev 12(5):1419–1434
7. Ferrer E (2012) A high order Discontinuous Galerkin-Fourier incompressible 3D Navier-Stokes solver with rotating sliding meshes for simulating cross-flow turbines. Ph.D. thesis, University of Oxford
8. Ferrer E, Willden RHJ (2011) A high order discontinuous Galerkin finite element solver for the incompressible Navier–Stokes equations. Comput Fluids 46(1):224–230
9. Ferrer E, Willden RHJ (2012) A high order discontinuous Galerkin-Fourier incompressible 3D Navier-Stokes solver with rotating sliding meshes. J Comput Phys 231(21):7037–7056
10. Ferrer E, Moxey D, Willden RHJ, Sherwin S (2014) Stability of projection methods for incompressible flows using high order pressure-velocity pairs of same degree: continuous and discontinuous galerkin formulations. Commun Comput Phys 16(3):817–840
11. Gretton GI, Bruce T (2005) Preliminary results from analytical and numerical models of a variable-pitch vertical-axis tidal current turbine. In: 6th European wave and tidal energy conference, Glasgow
12. Gretton GI, Bruce T (2006) Hydrodynamic modelling of a vertical-axis tidal current turbine using a Navier–Stokes solver. In: Proceedings of the 9th world renewable energy congress, Florence
13. Houlsby GT, Draper S, Oldefield MLG (2008) Application of linear momentum actuator disc theory to open channel flows. Technical report, Oxford University Engineering Library Report
14. Howell R, Qin N, Edwards J, Durrani N (2010) Wind tunnel and numerical study of a small vertical axis wind turbine. Renew Energ 35(2):412–422
15. McAdam R (2011) Studies into the technical feasibility of the transverse horizontal axis water turbine. Ph.D. thesis, University of Oxford
16. Newman BG (1983) Actuator-disc theory for vertical-axis wind turbines. J Wind Eng Ind Aerodyn 15(1–3):347–355
17. Paraschivoiu I (2002) Wind turbine design with emphasis on Darrieus concept. Polytechnic International Press, Montreal
18. Savage A (2007) Tidal power in the UK - tidal technologies overview, Technical Report Sustainable Developement Commission, 2007
19. Schlichting H (1979) Boundary layer theory. McGraw-Hill, New York
20. Strickland JH (1975) Darrieus turbine: a performance prediction model using multiple streamtubes. Technical report. Report No. SAND-75-0431, Sandia Labs
21. Templin RJ (1974) Aerodynamic performance theory for the NRC vertical-axis wind turbine. NASA STI/Recon Technical Report N, 76
22. Vermeer LJ, Sørensen JN, Crespo A (2003) Wind turbine wake aerodynamics. Prog Aerosp Sci 39(6–7):467–510
23. Whelan J, Thomson M, Graham JMR, Peiró J (2007) Modelling of free surface proximity and wave induced velocities around a horizontal axis tidal stream turbine. In: 7th European wave and tidal energy conference, Porto
24. Whelan JI, Graham JMR, Peiro J (2009) A free-surface and blockage correction for tidal turbines. J Fluid Mech 624:281–291
25. Wilson RE, Lissaman PBS (1974) Applied aerodynamics of wind power machines. NASA STI/Recon Technical report 75:22669

Chapter 2
Numerical Study of 2D Vertical Axis Wind and Tidal Turbines with a Degree-Adaptive Hybridizable Discontinuous Galerkin Method

Adeline Montlaur and Giorgio Giorgiani

Abstract This work presents a 2D study of vertical axis turbines with application to wind or tidal energy production. On the one hand, a degree-adaptive Hybridizable Discontinuous Galerkin (HDG) method is used to solve this incompressible Navier–Stokes problem. The HDG method allows to drastically reduce the coupled degrees of freedom (DOF) of the computation, seeking for an approximation of the solution that is defined only on the edges of the mesh. The discontinuous character of the solution provides an optimal framework for a degree-adaptive technique. Degree-adaptivity further reduces the number of DOF in the HDG computation by means of degree-refining only where more precision is needed. On the other hand, the finite volume method of ANSYS® is used to validate and compare the obtained results.

2.1 Introduction

In the last two decades, driven by the efforts to develop energy alternatives in order to reduce fossil fuel consumption and carbon emissions, wind power has been growing fast as an energy source, leading to cover 8 % of the European Union electricity consumption in 2013 [5]. Wind energy applications have first focused on locations where enhanced speed could be utilized to increase energy output, which explains why coastal and exposed hill tops have been preferred development sites, taking advantage of speed-up effects [2, 13]. More recently, especially in the last 5 years, offshore renewable energy, including tidal and offshore wind energy, has been undergoing rapid development globally. 2013 was indeed a record year for offshore installations, with 1.567 MW of new capacity grid connected, and with

A. Montlaur (✉)
Laboratori de Càlcul Numèric (LaCàN), Universitat Politècnica de Catalunya, Barcelona, Spain
e-mail: adeline.de.montlaur@upc.edu

G. Giorgiani
CNRS, Maison de la simulation, CEA Saclay, Gif-sur-Yvette, France
e-mail: giorgio.giorgiani@cea.fr

E. Ferrer, A. Montlaur (eds.), *CFD for Wind and Tidal Offshore Turbines*,
Springer Tracts in Mechanical Engineering, DOI 10.1007/978-3-319-16202-7_2

offshore wind power installations representing over 14 % of the annual EU wind energy market [5].

Since the technology is similar for onshore and offshore applications, offshore wind turbines are the most developed of the marine-based renewable energy resources. Lately, the prospect of increasing its share in electricity production has given rise to significant interest in modeling tidal turbines, since tidal current energy has the distinct advantage of being reliable and highly predictable. Though horizontal-axis turbines are the most classically used turbines onshore and offshore, vertical axis wind or tidal turbines are regarded as an alternative because of two attractive basic properties: they are cheaper in terms of installation and maintenance, since the rotor and the generator are installed at the base of the mast, and they are less sensitive to turbulence and to wind or water direction. There is also an ongoing offshore project on capitalizing on the energy potential available both in the winds above the ocean, and in the currents flowing beneath the waves. A prototype has been designed, by Mitsui Ocean Development Engineering Company [11], which uses an omnidirectional Darrieus wind turbine above the sea, on one end of a vertical shaft, along with a different type of omnidirectional turbine design, a Savonius one, spinning at the other end under the water surface.

In this context, special attention is needed on the analysis of new designs of vertical axis turbines with increased performance. For such a demanding technological application, numerical methods with higher accuracy than the standard first or second-order methods are an asset in order to allow a high-fidelity computation of the aerodynamic loads. Previous work (e.g., [6]) has considered high order DG to compute vertical axis turbines showing the necessity of high accuracy for this type of application. In this work, 2D vertical axis turbines are studied. On the one hand, a degree-adaptive Hybridizable Discontinuous Galerkin (HDG) method is used to solve this incompressible Navier–Stokes problem [3, 7, 8, 15]. The HDG method allows to drastically reduce the coupled degrees of freedom (DOF) of the computation, seeking for an approximation of the solution that is defined only on the edges of the mesh. The discontinuous character of the solution provides an optimal framework for a degree-adaptive (or p adaptation) technique, which further reduces the number of DOF in the HDG computation by means of degree-refining only where more precision is needed. This method is used along with a coordinate transformation that attaches the coordinate system to the rotating blades [10]. On the other hand, the second-order finite volume method of ANSYS® Academic Research CFD, Release 15.0 [1], is used to validate and complete the obtained results.

2.2 Variable Degree HDG for Incompressible Navier–Stokes

2.2.1 *Navier–Stokes Over a Broken Domain*

Let $\Omega \subset \mathbb{R}^d$ be an open bounded domain with boundary $\partial\Omega$ split in the Dirichlet, $\partial\Omega_D$, and Neumann, $\partial\Omega_N$, boundaries, and let T be the final instant of interest.

Recall the unsteady Navier–Stokes equations

$$\begin{cases} \partial_t \boldsymbol{u} + \nabla \cdot (\boldsymbol{u} \otimes \boldsymbol{u}) + \nabla p - \nu \Delta \boldsymbol{u} = \boldsymbol{f} & \text{in } \Omega \times]0, T[, \\ \nabla \cdot \boldsymbol{u} = 0 & \text{in } \Omega \times]0, T[, \\ \boldsymbol{u} = \boldsymbol{g} & \text{on } \partial\Omega_D \times]0, T[, \\ (-p\mathbf{I} + \nu \nabla \boldsymbol{u}) \cdot \boldsymbol{n} = \boldsymbol{t} & \text{on } \partial\Omega_N \times]0, T[, \\ \boldsymbol{u}(\boldsymbol{x}, 0) = \boldsymbol{u}_0 & \text{in } \Omega, \end{cases} \tag{2.1}$$

where $\boldsymbol{u}$ and p are the velocity and the kinematic pressure in the fluid, ν is the kinematic viscosity, $\boldsymbol{f}$ is a body force, $\boldsymbol{n}$ is the unitary outward normal vector, $\mathbf{I}$ is the identity matrix, $\boldsymbol{g}$ is the prescribed velocity on the Dirichlet boundary $\partial\Omega_D$, $\boldsymbol{t}$ are the prescribed pseudo-tractions imposed on the Neumann boundary $\partial\Omega_N$, and $\boldsymbol{u}_0$ is the initial velocity field (assumed solenoidal: $\nabla \cdot \boldsymbol{u}_0 = 0$).

Note that these Neumann boundary conditions do not correspond to regular stresses but to pseudo-stresses, as it is standard in velocity–pressure formulation, see [4]. However, more *physical* stress boundary conditions can also be implemented, see [7] for a detailed description.

For discontinuous Galerkin approaches the domain Ω is partitioned in $\mathtt{n_{el}}$ disjoint elements Ω_i with boundaries $\partial\Omega_i$, such that

$$\overline{\Omega} = \bigcup_{i=1}^{\mathtt{n_{el}}} \overline{\Omega}_i, \quad \Omega_i \cap \Omega_j = \emptyset \text{ for } i \neq j,$$

and the union of all $\mathtt{n_{fc}}$ faces (sides for 2D) is denoted as

$$\Gamma := \bigcup_{i=1}^{\mathtt{n_{el}}} \partial\Omega_i .$$

The discontinuous setting induces a new problem equivalent to (2.1). It is written as a system of first-order partial differential equations (mixed form) with some element-by-element equations and some global ones, namely, for $i = 1, \ldots, \mathtt{n_{el}}$

$$\left.\begin{aligned} \boldsymbol{L} - \nabla \boldsymbol{u} &= \boldsymbol{0} \\ \partial_t \boldsymbol{u} + \nabla \cdot (\boldsymbol{u} \otimes \boldsymbol{u} + p\mathbf{I} - \nu \boldsymbol{L}) &= \boldsymbol{f} \\ \nabla \cdot \boldsymbol{u} &= 0 \end{aligned}\right\} \quad \text{in } \Omega_i \times]0, T[, \tag{2.2a}$$

$$\boldsymbol{u}(\boldsymbol{x}, 0) = \boldsymbol{u}_0 \quad \text{in } \Omega_i, \tag{2.2b}$$

and

$$\llbracket \boldsymbol{u} \otimes \boldsymbol{n} \rrbracket = \boldsymbol{0} \qquad \text{on } \Gamma \backslash \partial\Omega \times]0, T[, \tag{2.2c}$$

$$\llbracket (-p\mathbf{I} + \nu \boldsymbol{L}) \cdot \boldsymbol{n} \rrbracket = \boldsymbol{0} \qquad \text{on } \Gamma \backslash \partial\Omega \times]0, T[, \tag{2.2d}$$

$$\boldsymbol{u} = \boldsymbol{g} \qquad \text{on } \partial\Omega_D \times]0, T[, \tag{2.2e}$$

$$(-p\mathbf{I} + \nu \boldsymbol{L}) \cdot \boldsymbol{n} = \boldsymbol{t} \qquad \text{on } \partial\Omega_N \times]0, T[, \tag{2.2f}$$

where a new variable $\boldsymbol{L}$ for the velocity gradient tensor is introduced after splitting the second-order momentum conservation equation in two first-order equations. The *jump* $\llbracket \cdot \rrbracket$ operator is defined at each internal face of Γ, i.e. on $\Gamma \backslash \partial\Omega$, using values from the elements to the left and right of the face (say, Ω_i and Ω_j), namely

$$\llbracket \circledcirc \rrbracket = \circledcirc_i + \circledcirc_j ,$$

and always involving the normal vector $\boldsymbol{n}$, see [12] for details. Thus, Eq. (2.2c) imposes the continuity of velocity and Eq. (2.2d) imposes the continuity of the normal component of the pseudo-stress across interior faces.

2.2.2 The HDG Local Problem

A major feature of HDG is that, in general, unknowns are restricted to the skeleton of the mesh, that is, the union of all faces denoted by Γ. Here, the velocity field, $\hat{\boldsymbol{u}}(\boldsymbol{x}, t)$, on the mesh skeleton Γ is this unknown. However, as described next, satisfying the incompressibility equation requires the introduction of one scalar unknown per element, irrespective of the polynomial degree used in the approximation. There are many choices for this scalar unknown; here, the mean pressure on the element boundary is chosen. The introduction of the new variable $\hat{\boldsymbol{u}}(\boldsymbol{x}, t)$ on the mesh skeleton Γ is crucial to define two types of problems: a local problem for each element and a global one for all faces.

In fact, the local element-by-element problem corresponds to the Navier–Stokes equations on each element, see Eq. (2.2a), with imposed Dirichlet boundary conditions. The imposed Dirichlet conditions on the element boundary are precisely the velocities $\hat{\boldsymbol{u}}(\boldsymbol{x}, t)$ for $\boldsymbol{x} \in \Gamma$.

It is well known that an incompressible Navier–Stokes problem on a bounded domain with non-homogeneous velocity prescribed everywhere on the boundary requires a solvability condition and implies that pressure is known up to a constant. More precisely, the continuity equation implies a zero flux compatibility condition. Thus, the newly introduced variable $\hat{\boldsymbol{u}}(\boldsymbol{x}, t)$ on the mesh skeleton Γ must verify

$$\langle \hat{\boldsymbol{u}} \cdot \boldsymbol{n}, 1 \rangle_{\partial\Omega_i} = 0, \quad \text{for } i = 1, \ldots, \mathtt{n_{el}}. \tag{2.3}$$

Now the local element-by-element Navier–Stokes equations can be solved to determine $(\boldsymbol{u}, \boldsymbol{L}, p)$ in terms of the imposed $\hat{\boldsymbol{u}}(\boldsymbol{x}, t)$ on the mesh skeleton Γ. Thus, for $i = 1, \dots, \mathtt{n_{el}}$ the local HDG problem is solved, namely

$$\left.\begin{aligned} \boldsymbol{L} - \boldsymbol{\nabla}\boldsymbol{u} &= \boldsymbol{0} \\ \partial_t \boldsymbol{u} + \boldsymbol{\nabla}\cdot(\boldsymbol{u}\otimes\boldsymbol{u} + p\mathbf{I} - \nu\boldsymbol{L}) &= \boldsymbol{f} \\ \boldsymbol{\nabla}\cdot\boldsymbol{u} &= 0 \end{aligned}\right\} \qquad \text{in } \Omega_i \times]0, T[, \tag{2.4a}$$

$$\boldsymbol{u}(\boldsymbol{x}, 0) = \boldsymbol{u}_0 \qquad \text{in } \Omega_i, \tag{2.4b}$$

$$\boldsymbol{u} = \hat{\boldsymbol{u}} \qquad \text{on } \partial\Omega_i \times]0, T[, \tag{2.4c}$$

$$\langle p, 1\rangle_{\partial\Omega_i} = \rho_i. \tag{2.4d}$$

As noted earlier, (2.4) is a Dirichlet problem and, consequently, pressure is determined up to a constant. This constant is determined by prescribing some value for the pressure. Typical choices are to impose pressure at one point or to prescribe the mean value in the domain. In HDG, as noted earlier and since the unknowns are restricted to the mesh skeleton, the usual choice is to prescribe the mean pressure on the element boundary, see (2.4d), where $\langle\cdot,\cdot\rangle_B$ denotes the $\mathcal{L}^2$ scalar product of the traces over any $B \subset \Gamma$.

At this point it is important to notice that given the values of the velocities on Γ, $\hat{\boldsymbol{u}}(\cdot, t) \in [\mathcal{L}^2(\Gamma)]^d$ for any instant $t \in [0, T]$, the same Dirichlet boundary condition is imposed to the left and right element of a given face. Consequently, the velocity continuity, recall Eq. (2.2c), is ensured by (2.4c). Obviously, on the Dirichlet boundary $\hat{\boldsymbol{u}} = \boldsymbol{g}$, replicating (2.2e).

Observe that in each element the original unknowns, $(\boldsymbol{L}, \boldsymbol{u}, p)$, can be determined in terms of the two extra unknowns: the velocity on Γ, $\hat{\boldsymbol{u}}$, and the vector of average pressures on the boundaries for each element, $\{\rho\}_{i=1}^{\mathtt{n_{el}}} \in \mathbb{R}^{\mathtt{n_{el}}}$.

An approximation is obtained after the corresponding discretization, see [15]. Two types of finite dimensional spaces must be defined, one for functions in the elements interior and the other for trace functions, namely

$$\begin{aligned} \mathcal{V}_t^h &:= \{v : v(\cdot, t) \in \mathcal{V}^h \text{ for any } t \in [0, T]\}, \text{ with} \\ \mathcal{V}^h &:= \{v \in \mathcal{L}^2(\Omega) : v|_{\Omega_i} \in \mathcal{P}^{k_{\Omega_i}}(\Omega_i) \text{ for } i = 1, \dots, \mathtt{n_{el}}\}, \text{ and} \\ \Lambda_t^h &:= \{\hat{v} : \hat{v}(\cdot, t) \in \Lambda^h \text{ for any } t \in [0, T]\}, \text{ with} \\ \Lambda^h &:= \{\hat{v} \in \mathcal{L}^2(\Gamma) : \hat{v}|_{\Gamma_i} \in \mathcal{P}^{k_{\Gamma_i}}(\Gamma_i) \text{ for } i = 1, \dots, \mathtt{n_{fc}}\}, \end{aligned}$$

where $\mathcal{P}^k$ denotes the space of polynomials of degree less than or equal to k, while k_{Ω_i} and k_{Γ_i} are the polynomial degrees in element Ω_i and face Γ_i, respectively. To simplify the presentation, in an abuse of notation, the same notation is used for the numerical approximation, belonging to the finite dimensional spaces, and the exact solution, that is $(u, \boldsymbol{L}, p)$.

The weak problem for each element corresponding to (2.4) becomes: given $\{\rho_i\}_{i=1}^{\mathtt{n_{el}}} \in \mathbb{R}^{\mathtt{n_{el}}}$ and $\hat{\boldsymbol{u}} \in [\Lambda_t^h]^d$ satisfying (2.3), find an approximation $(\boldsymbol{L}, \boldsymbol{u}, p) \in [\mathcal{V}_t^h]^{d\times d} \times [\mathcal{V}_t^h]^d \times \mathcal{V}_t^h$ such that

$$\begin{aligned}
\left(\boldsymbol{G}, \boldsymbol{L}\right)_{\Omega_i} + \left(\nabla\cdot\boldsymbol{G}, \boldsymbol{u}\right)_{\Omega_i} - \langle\boldsymbol{G}\cdot\boldsymbol{n}, \hat{\boldsymbol{u}}\rangle_{\partial\Omega_i} &= 0,\\
\left(\boldsymbol{v}, \partial_t\boldsymbol{u}\right)_{\Omega_i} + \left(\nu\boldsymbol{L} - p\mathbf{I} - \boldsymbol{u}\otimes\boldsymbol{u}, \nabla\boldsymbol{v}\right)_{\Omega_i} &\\
+\langle(-\nu\boldsymbol{L} + p\mathbf{I} + \hat{\boldsymbol{u}}\otimes\hat{\boldsymbol{u}})\cdot\boldsymbol{n} + \tau(\boldsymbol{u}-\hat{\boldsymbol{u}}), \boldsymbol{v}\rangle_{\partial\Omega_i} &= \left(\boldsymbol{v}, \boldsymbol{f}\right)_{\Omega_i},\\
-\left(\nabla q, \boldsymbol{u}\right)_{\Omega_i} + \langle q, \hat{\boldsymbol{u}}\cdot\boldsymbol{n}\rangle_{\partial\Omega_i} &= 0,\\
\langle p, 1\rangle_{\partial\Omega_i} &= \rho_i,
\end{aligned} \tag{2.5}$$

for all $(\boldsymbol{G}, \boldsymbol{v}, q) \in [\mathcal{V}^h]^{d\times d} \times [\mathcal{V}^h]^d \times \mathcal{V}^h$, for $i = 1, \ldots, \mathtt{n_{el}}$, and with the initial condition defined in (2.4b). As usual, $(\cdot,\cdot)_{\Omega_i}$ denotes the $\mathcal{L}^2$ scalar product in the element Ω_i.

In this weak problem it is important to note two details. First, the Dirichlet boundary conditions, (2.4c), are imposed weakly; and second, the trace of the normal stress has been replaced in all boundary integrals by the following numerical trace

$$(-p\widehat{\mathbf{I} + \nu}\boldsymbol{L})\cdot\boldsymbol{n} := (-p\mathbf{I} + \nu\boldsymbol{L})\cdot\boldsymbol{n} + \tau(\hat{\boldsymbol{u}} - \boldsymbol{u}), \tag{2.6}$$

where τ is a stability parameter. A discussion on the choice of the stabilization parameter for a degree-adaptive HDG method applied to the Navier–Stokes equations can be found in [7]. A constant τ in the whole domain is suggested, with value $\tau \approx \|\boldsymbol{u}\|$, where $\|\boldsymbol{u}\|$ is a characteristic velocity. Here, the problem involves a moving reference frame, which is rotating with the turbine. Thus, the body forces appearing in the computational domain generate a rotating velocity whose magnitude increases with the distance from the rotation center. For this reason, particular care must be taken with the choice of the stabilization parameter τ. In fact, numerical evidence shows that if a constant τ is used, either loss of stability or loss of superconvergence may deteriorate the numerical solution. Hence, a variable τ is used in this case, with the following relation

$$\tau = \begin{cases} \tau_0\omega\|\boldsymbol{x}\| & \text{if } \omega\|\boldsymbol{x}\| \geq 1,\\ \tau_0 & \text{if } \omega\|\boldsymbol{x}\| < 1, \end{cases}$$

where $\boldsymbol{x}$ is the position vector pointing at the barycenter of the face on which τ is defined, ω is the angular velocity of the turbine (assumed positive), and τ_0 is a constant value for the whole mesh.

2.2.3 The HDG Global Problem

The local problems (2.4), or (2.5), allow to compute the solution in the whole domain (that is, the variables $\boldsymbol{L}, \boldsymbol{u}$ and p) in terms of the trace of the velocity on the mesh skeleton, $\hat{\boldsymbol{u}}$, and the mean pressure for each element, $\{\rho_i\}_{i=1}^{\mathrm{n_{el}}}$. Thus, $\hat{\boldsymbol{u}}$ and ρ_i can now be understood as the actual unknowns of the problem. They are determined using the global equations in (2.2), in particular (2.2d) and (2.2f), and the solvability condition (2.3). In fact, as already discussed, Eqs. (2.2c) and (2.2e) are already imposed. Thus (2.2d) and (2.2f) are the remaining global conditions which must be imposed. These two equations (in weak form) and the solvability condition (2.3) determine the HDG *global problem*. Namely, find approximations $\hat{\boldsymbol{u}} \in [\Lambda_t^h(\boldsymbol{g})]^d$ and $\{\rho_i\}_{i=1}^{\mathrm{n_{el}}} \in \mathbb{R}^{\mathrm{n_{el}}}$ such that

$$\sum_{i=1}^{\mathrm{n_{el}}} \big\langle \hat{\boldsymbol{v}}, (-p\mathbf{I} + \nu \boldsymbol{L}) \cdot \boldsymbol{n} + \tau(\hat{\boldsymbol{u}} - \boldsymbol{u}) \big\rangle_{\partial\Omega_i} = \big\langle \hat{\boldsymbol{v}}, \boldsymbol{t} \big\rangle_{\partial\Omega_N}, \tag{2.7a}$$

$$\big\langle \hat{\boldsymbol{u}} \cdot \boldsymbol{n}, 1 \big\rangle_{\partial\Omega_i} = 0, \text{ for } i = 1, \ldots, \mathrm{n_{el}}, \tag{2.7b}$$

for all $\hat{\boldsymbol{v}} \in [\Lambda^h(\mathbf{0})]^d$. Here $(\boldsymbol{L}, \boldsymbol{u}, p) \in [\mathcal{V}_t^h]^{d\times d} \times [\mathcal{V}_t^h]^d \times \mathcal{V}_t^h$ are solution of the local problems (2.5) and the trace spaces associated to the Dirichlet boundary are defined by $[\Lambda^h(\square)]^d = \{\hat{\boldsymbol{v}} \in [\Lambda^h]^d : \hat{\boldsymbol{v}} = \mathbb{P}_\partial \square \text{ on } \partial\Omega_D\}$, with $\mathbb{P}_\partial$ the $\mathcal{L}^2$ projection on $\partial\Omega_D$.

Note that Eq. (2.2d) imposes continuity of the normal component of the pseudo-traction on each element face, which induces (2.7a) after using (2.6). Thus Eq. (2.7a) weakly imposes the continuity of the normal pseudo-stress.

Remark 1 (Nonlinear DAE System) The HDG discrete problem is defined by (2.5) and (2.7). It is a system of Differential Algebraic Equations (DAE) of index 1, that can be efficiently discretized in time with an implicit time integrator, such as backward Euler, a Backward Differentiation Formula, or a diagonally implicit Runge–Kutta method, see [14, 15]. Time discretization of (2.5) and (2.7) leads to a non-linear system of equations that can be solved with an iterative scheme. Here, the non-linear system has been linearized using the Newton–Raphson method. That is, every non-linear convective term in (2.5), which can be expressed as a trilinear form $c(\boldsymbol{u}, \boldsymbol{u}; \boldsymbol{v})$, is linearized using the first-order approximation $c(\boldsymbol{u}^r, \boldsymbol{u}^r; \boldsymbol{v}) \approx c(\boldsymbol{u}^{r-1}, \boldsymbol{u}^r; \boldsymbol{v}) + c(\boldsymbol{u}^r, \boldsymbol{u}^{r-1}; \boldsymbol{v}) - c(\boldsymbol{u}^{r-1}, \boldsymbol{u}^{r-1}; \boldsymbol{v})$, where here r is the iteration count. Obviously, this implies that the global solution, Eq. (2.7), must be iterated.

In any case, a linear system of equations is solved for each iteration of the non-linear solver. In this linear system, the equations corresponding to (2.5) can be solved element-by-element to express the solution at each element Ω_i in terms of the trace variable, $\hat{\boldsymbol{u}}$, and the mean of the pressure in the element boundary, ρ_i. Then, these expressions are replaced in (2.7) yielding a global system of equations that only involves $\hat{\boldsymbol{u}}$ and $\{\rho\}_{i=1}^{\mathrm{n_{el}}}$, with an important reduction in number of DOF. Further details on the efficient solution of the non-linear DAE can be found in [15].

Remark 2 (HDG Post-processed Solution) The solution of the HDG problem, given by Eqs. (2.5) and (2.7), provides a numerical solution $(\boldsymbol{L}, \boldsymbol{u}) \in [\mathcal{V}_t^h]^{d\times d} \times [\mathcal{V}_t^h]^d$ with optimal numerical convergence in both variables. Then, a new problem can be solved element-by-element to compute a superconvergent approximation of the velocity $\boldsymbol{u}^*$, namely, for $i = 1, \ldots, \mathtt{n_{el}}$, solve

$$\begin{cases} \nabla \cdot \nabla \boldsymbol{u}^* = \nabla \cdot \boldsymbol{L} & \text{in } \Omega_i, \\ \boldsymbol{n} \cdot \nabla \boldsymbol{u}^* = \boldsymbol{n} \cdot \boldsymbol{L} & \text{on } \partial\Omega_i, \\ \left(\boldsymbol{u}^*, 1\right)_{\Omega_i} = \left(\boldsymbol{u}, 1\right)_{\Omega_i}. \end{cases}$$

This induces a weak problem in a richer finite dimensional space, that is, find $\boldsymbol{u}^* \in [\mathcal{V}_t^{h*}]^d$ such that

$$\left(\nabla \boldsymbol{v}, \nabla \boldsymbol{u}^*\right)_{\Omega_i} = \left(\nabla \boldsymbol{v}, \boldsymbol{L}\right)_{\Omega_i} \text{ and } \left(\boldsymbol{u}^*, 1\right)_{\Omega_i} = \left(\boldsymbol{u}, 1\right)_{\Omega_i},$$

for all $\boldsymbol{v} \in [\mathcal{V}^{h*}]^d$ and $i = 1, \ldots, \mathtt{n_{el}}$, where $\mathcal{V}^{h*}$ must be a bigger space than $\mathcal{V}^h$. In fact, with one degree more in the element-by-element polynomial approximation, i.e. $\mathcal{V}^{h*} = \left\{v \in \mathcal{L}^2(\Omega) : v|_{\Omega_i} \in \mathcal{P}^{k_{\Omega_i}+1}(\Omega_i), \text{ for } i = 1, \ldots, \mathtt{n_{el}}\right\}$, $\boldsymbol{u}^*$ converges asymptotically at a rate $k+2$ in the $\mathcal{L}^2$ norm. More details on the superconvergence properties of the post-processed solution, in particular in the case of non-uniform degree distribution, can be found in [7]. Note that the post-process solution is not required to be computed at each time step, but only when an improved solution is needed. Moreover, the computational overhead is small and decreases with the approximation degree, see [9].

In the following section, the superconvergent solution $\boldsymbol{u}^*$ is used to compute a reliable and inexpensive error estimator for the HDG velocity approximation $\boldsymbol{u}$.

2.3 Error Estimation and Degree-Adaptive Algorithm

In the next section, the adaptive process introduced in [7] is briefly described and used in the degree-adaptive HDG simulation of the vertical axis turbine. It is based on an inexpensive, reliable, and computable error estimator for the velocity field, derived using the superconvergent HDG post-processed solution.

More precisely, the $\mathcal{L}_2$ error in the velocity field $\boldsymbol{u}$ is estimated in an element Ω_i as

$$E_i^2 = \frac{1}{|\Omega_i|} \int_{\Omega_i} (\boldsymbol{u}^* - \boldsymbol{u})^2 \, d\Omega, \tag{2.8}$$

where $\boldsymbol{u}$ is the solution of the HDG problem, see Eqs. (2.5) and (2.7), and $\boldsymbol{u}^*$ is the improved velocity, see Remark 2. The elemental error is normalized by the element measure: this is crucial for non-uniform meshes, see [7].

Given the error estimate (2.8), an automatic degree-adaptive process is proposed. A tolerance ε for the velocity elemental error is assumed. The adaptive process aims at getting a map of elemental degrees $\{k_{\Omega_i}\}_{i=1}^{\mathtt{n_{el}}}$ such that the error estimate of the HDG solution satisfies

$$E_i \leq \varepsilon \quad \forall\, \Omega_i \subset \Omega. \tag{2.9}$$

Hence using the error distribution, a map of degree increments Δk_{Ω_i} is evaluated for each element in the whole computational domain Ω. The degree variation in each element is computed as

$$\Delta k_{\Omega_i} = \left\lceil \log_b (E_i/\varepsilon) \right\rceil, \text{ for } i = 1, \ldots, \mathtt{n_{el}}. \tag{2.10}$$

where $\lceil \cdot \rceil$ denotes the ceiling function and b is the so-called *adaptation aggressiveness*, see [7].

The adaption process starts with a uniform degree mesh. The error estimation and consequent degree adaptation is performed after a fixed number of time steps N. That is, every N time steps, the superconvergent velocity $\boldsymbol{u}^*$ is computed, the error estimate (2.8) is evaluated, the degree map is updated using (2.10), the solution is projected onto the new computational mesh, and the time integration is continued.

The flow chart of the adaptive strategy is shown in Fig. 2.1.

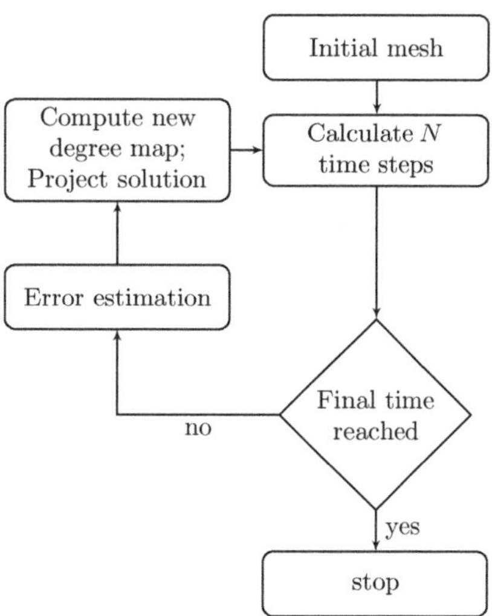

Fig. 2.1 Flow chart of the adaptive strategy

Numerical experiments in [7] show that, even though the adaptive process is based on an error estimate for the velocity field, accurate approximations of the fluid-dynamic forces are also obtained.

2.4 Numerical Simulation of a Vertical Axis Turbine

2.4.1 Moving Reference Frame

For the application of a vertical axis turbine, which is a body moving with a rotational angle $\theta = \theta(t) = \omega t$, where ω is the angular velocity of the turbine, in an absolute frame of reference $\boldsymbol{x}' = (x', y')$, a corresponding moving frame of reference can be attached on the body with the transformation

$$x' = x\cos\theta + y\sin\theta \tag{2.11}$$

$$y' = -x\sin\theta + y\cos\theta \tag{2.12}$$

the transformation of velocity is thus

$$\boldsymbol{u}' = \mathbf{A}(\boldsymbol{u} - \dot{\theta}\mathbf{I}_0\boldsymbol{x}) \tag{2.13}$$

with

$$\mathbf{A} = \begin{pmatrix} \cos\theta & \sin\theta \\ -\sin\theta & \cos\theta \end{pmatrix} \text{ and } \mathbf{I}_0 = \begin{pmatrix} 0 & -1 \\ 1 & 0 \end{pmatrix}. \tag{2.14}$$

System (2.1) is thus modified as follows for the momentum equation

$$\partial_t \boldsymbol{u} + \nabla\cdot(\boldsymbol{u}\otimes\boldsymbol{u}) + \nabla p - \nu\Delta\boldsymbol{u} = \boldsymbol{f} + 2\dot{\theta}\mathbf{I}_0\boldsymbol{u} + (\dot{\theta}^2 + \ddot{\theta}\mathbf{I}_0)\boldsymbol{x} \quad \text{in } \Omega, \tag{2.15}$$

and Eq. (2.13) is used to transform Dirichlet boundary condition. Finally, for the Neumann boundary conditions results

$$(-p\mathbf{I} + \nu\nabla\boldsymbol{u})\boldsymbol{n} = \dot{\theta}\mathbf{I}_0\boldsymbol{n} \quad \text{on } \partial\Omega_N.$$

More details can be found in [10].

2.4.2 Hypothesis

A dimensionless problem consisting of a 2D 4-blade vertical axis turbine is considered. The blades consist of NACA 0012 airfoils with a chord of dimension

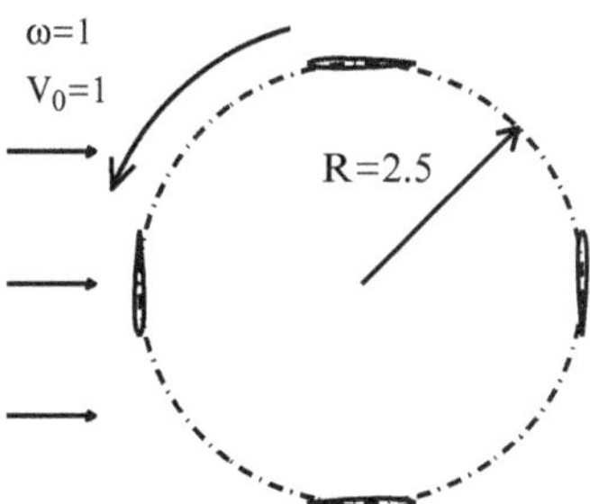

Fig. 2.2 Geometry and setup of the 4-blade vertical axis turbine

1.2, and the radius of the turbine is 2.5. An angular velocity $\omega = 1$ for the turbine and an upstream velocity $V_0 = 1$ are set. The Reynolds number is $\text{Re} = 1/\nu = 1000$. The statement of the problem is shown in Fig. 2.2.

Numerical simulations were carried out using the degree-adaptive HDG method described in Sect. 2.2 and with the commercial CFD solver, ANSYS® Academic Research CFD, Release 15.0 [1], using the sliding mesh model from the Fluent solver.

An unstructured triangular mesh composed by 4,474 elements was used for the HDG method. The curved boundaries of the NACA profiles were represented by piecewise second-order polynomials . The adaptive HDG procedure was set up using a parameter $b = 10$ and a tolerance $\varepsilon = 10^{-4}$, and the error estimation and adaptive procedure was performed every 10 time steps.

In the Fluent solver, all velocity calculations use a second-order bounded differencing scheme, and a second-order scheme is used for all diffusive terms. The SIMPLE algorithm, using a relationship between velocity and pressure corrections to enforce mass conservation and to obtain the pressure field, is applied. To compare adequately the results, no turbulence model is set in the Fluent solver. The ANSYS® computational mesh is composed of 24,786 quadrilateral elements. A body sizing of 0.03 is used in the region of interest, that is, in the turbine vicinity. Within the computational mesh, the coarsest element size was set to 0.5. Both methods use a first-order implicit time integration, with a time step of $\Delta t = 0.01$.

2.4.3 Results

Figure 2.3 illustrates how adaptive computation is particularly suited for moving geometries and reference frames. Figure 2.3a shows the vorticity in the area surrounding the turbine blades, whereas Fig. 2.3b shows how the adaptive technique correctly places high-order elements around the blades and in the generated wake, where more precision is needed to properly capture the flow properties.

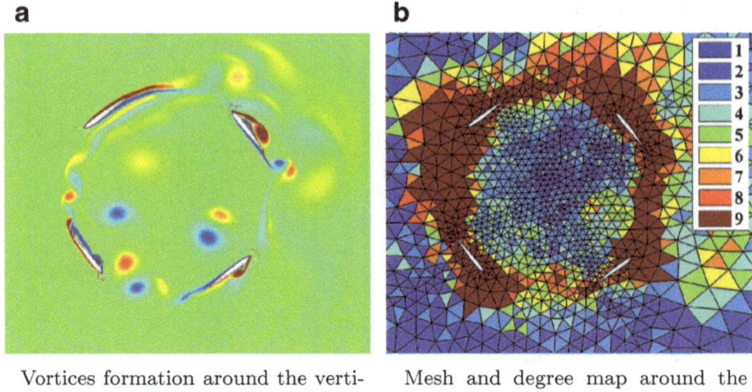

Vortices formation around the vertical axis turbine

Mesh and degree map around the turbine.

Fig. 2.3 Vorticity (**a**) and degree map (**b**) for $t = 4.3$

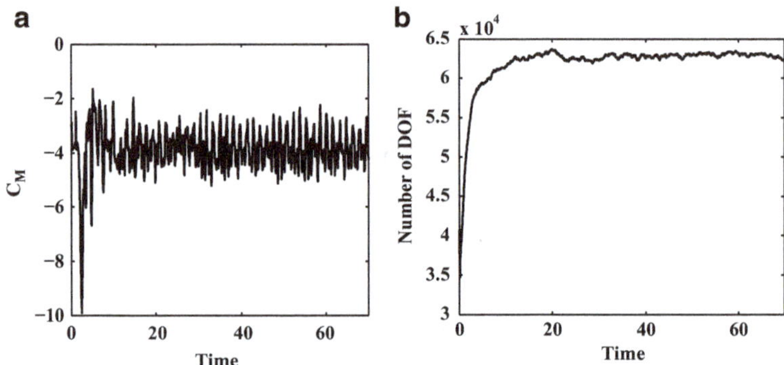

Fig. 2.4 Moment coefficient (**a**) and number of DOF (**b**) as functions of time evolution of the simulation

Figure 2.4 shows the time evolutions of the moment coefficient C_M of the turbine and of the number of DOF. As expected, after the initial transient where the degree-adaptive algorithm increases the degree of interpolation when needed, the number of DOF reaches an almost constant value when the periodic state is achieved.

Figure 2.5 shows a frequency analysis of the lift and drag coefficients of the turbine. A good agreement between the HDG solution and the one computed with the commercial code ANSYS® is found, and the main harmonics are well captured by both methods. Note that the good correlation of the results observed is obtained at similar cost as Table 2.1 shows. HDG allows using a coarser mesh by increasing the degree of interpolation up to 9 in the region where more precision is needed, see also Fig. 2.3b, whereas ANSYS® needs a finer mesh since the interpolation degree is the same everywhere. This leads to a similar cost, in terms of number of DOF, for both methods.

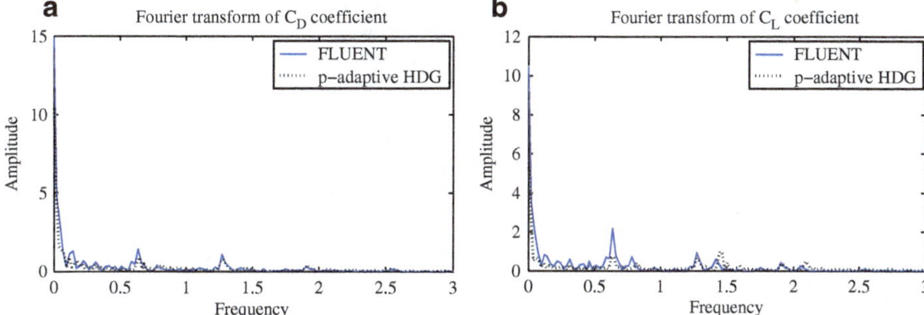

Fig. 2.5 Frequency analysis of the drag (**a**) and lift (**b**) coefficient of the vertical turbine obtained with the HDG method and with ANSYS®

Table 2.1 Cost and degree comparison between degree-adaptive HDG method and ANSYS-Fluent

	HDG	Fluent
Number of elements	4,474	24,786
Number of DOF	60,000	75,000
Degree of interpolation	1–9	0

2.5 Conclusions

This study shows the advantage of a degree-adaptive technique for moving reference frame applications. The HDG method is particularly suited for implanting this technique due to, on the one hand, the discontinuous character of the solution that allows to easily handle variable degree meshes. On the other hand, the superconvergence properties of the HDG solution allow to derive a simple, reliable, and computationally cheap error estimator to drive the automatic adaptive process. With a simple strategy that does not involve modification of the mesh, the proposed degree-adaptive technique relaxes the need of a priori designing finite element discretization to correctly resolve the solution.

An example of a 2D cross-section of a 4-blade vertical axis turbine is considered in order to validate the method and compare the results with the commercial software ANSYS®. Similar results are obtained at a very competitive cost for the HDG method. Future work should incorporate a turbulence model, which is indispensable in this type of studies.

References

1. ANSYS 15.0 Product Documentation (2013)
2. Cochard S, Letchford CW, Earl TA, Montlaur A (2012) Formation of tip-vortices on triangular prismatic-shaped cliffs. Part 1: A wind tunnel study. J Wind Eng Ind Aerodyn 109:9–20
3. Cockburn B, Nguyen NC, Peraire J (2010) A comparison of HDG methods for Stokes flow. J Sci Comput 45(1–3):215–237

4. Donea J, Huerta A (2003) Finite element methods for flow problems. Wiley, Chichester
5. European Wind Energy Association (2014) Wind in power: 2013 European Statistics
6. Ferrer E, Willden R (2012) A high order discontinuous Galerkin-Fourier incompressible 3D Navier-Stokes solver with rotating sliding meshes. J Comput Phys 231(21):7037–7056
7. Giorgiani G, Fernández-Méndez S, Huerta A (2014) Hybridizable discontinuous Galerkin with degree adaptivity for the incompressible Navier-Stokes equations. Comput Fluids 98:196–208
8. Giorgiani G, Fernández-Méndez S, Huerta A (2013) Hybridizable discontinuous Galerkin p-adaptivity for wave propagation problems. Int J Numer Methods Fluids 72(12):1244–1262
9. Giorgiani G, Fernández-Méndez S, Huerta A (2013) High-order continuous and discontinuous Galerkin methods for wave. Int J Numer Methods Fluids 73(10):883–903
10. Li L, Sherwin SJ, Bearman PW (2002) A moving frame of reference algorithm for fluid/structure interaction of rotating and translating bodies. Int J Numer Methods Fluids 38(2):187–206
11. MODEC (2013) Floating wind and current hybrid power generation—Savonius Keel and wind turbine Darrieus. http://www.modec.com/fps/skwid/pdf/skwid.pdf. Accessed 02 Dec 2014
12. Montlaur A, Fernandez-Mendez S, Huerta A (2008) Discontinuous Galerkin methods for the Stokes equations using divergence-free approximations. Int J Numer Methods Fluids 57(9):1071–1092
13. Montlaur A, Cochard S, Fletcher DF (2012) Formation of tip-vortices on triangular prismatic-shaped cliffs. Part 2: A computational fluid dynamics study. J Wind Eng Ind Aerodyn 109:21–30
14. Montlaur A, Fernandez-Mendez S, Huerta A (2012) High-order implicit time integration for unsteady incompressible flows. Int J Numer Methods Fluids 70(5):603–626
15. Nguyen NC, Peraire J, Cockburn B (2011) An implicit high-order hybridizable discontinuous Galerkin method for the incompressible Navier-Stokes equations. J Comput Phys 230(4):1147–1170

Chapter 3
A Moving Least Squares-Based High-Order-Preserving Sliding Mesh Technique with No Intersections

Luis Ramírez, Xesús Nogueira, Charles Foulquié, Sofiane Khelladi, Jean-Camille Chassaing, and Ignasi Colominas

Abstract The sliding mesh approach is widely used in numerical simulation of turbomachinery flows to take in to account the rotor/stator or rotor/rotor interaction. This technique allows relative sliding of one grid adjacent to another grid (static or in motion). However, when a high-order method is used, the interpolation used in the sliding mesh model needs to be of, at least, the same order than the numerical scheme, in order to prevent loss of accuracy. In this work we present a sliding mesh model based on the use of Moving Least Squares (MLS) approximations. It is used with a high-order (>2) finite volume method that computes the derivatives of the Taylor reconstruction inside each control volume using MLS approximants. Thus, this new sliding mesh model fits naturally in a high-order MLS-based finite volume framework (Cueto-Felgueroso et al., Comput Methods Appl Mech Eng 196:4712–4736, 2007; Khelladi et al., Comput Methods Appl Mech Eng 200:2348–2362, 2011) for the computation of acoustic wave propagation into turbomachinery.

3.1 Introduction

Renewable sources of energy, such as wind and tidal power are expected to play a key role in the substitution of conventional energy sources. Wind energy generation systems have been greatly developed in the last two decades. Thus, current wind turbine designs are near their technical limit and new paradigms of design have to

L. Ramírez • X. Nogueira (✉) • I. Colominas
Group of Numerical Methods in Engineering, Universidade da Coruña, Campus de Elviña, 15071 A Coruña, Spain
e-mail: xnogueira@udc.es

S. Khelladi • C. Foulquié
Arts et Métiers Paris Tech, DynFluid Lab, 151 Boulevard de l'Hôpital, 75013 Paris, France

J.-C. Chassaing
Sorbonne Universités, UPMC Univ. Paris 06, CNRS, UMR7190, D'Alembert Institute, 75005 Paris, France

E. Ferrer, A. Montlaur (eds.), *CFD for Wind and Tidal Offshore Turbines*, Springer Tracts in Mechanical Engineering, DOI 10.1007/978-3-319-16202-7_3 be found. The case of tidal energy generation is the opposite, since the technology of tidal turbines is still in its infancy. In this context, CFD appears as a powerful tool in the design of new prototypes.

In the numerical simulation of turbomachinery flows, a problem appears when it is required to account for the relative motion between rotor and stator or the interaction between different blades of the rotor. One numerical technique widely used to take into account the rotor/stator or rotor/rotor interaction flow is the so-called sliding mesh [10, 16] method. This technique allows relative sliding of one grid adjacent to another grid (static or in motion). Thus, non-matching cells [14, 15] may appear at the interface between static and moving grids. This introduces a problem of interpolation. In addition, when a high-order method is used, the interpolation used in the sliding mesh model needs to be of, at least, the same order than the numerical scheme, in order to prevent loss of accuracy. In this context, a higher-order ($\geq$3) discontinuous Galerkin method with sliding mesh capabilities was proposed by Ferrer and Willden [5].

Some of the authors of this work have proposed a high-resolution finite volume method based on Moving Least Squares (MLS) reconstructions. The theoretical fundamentals of this finite volume method (FV-MLS) were presented in [3, 6, 11–13] and the references therein. In this work we present a sliding mesh model based on the use of MLS approximations [7]. This model is used with a high-order (>2) finite volume method that computes the derivatives of the Taylor reconstruction inside each control volume using MLS approximations [3, 6, 12, 13]. Thus, this new sliding mesh model fits naturally in a high-order finite volume framework.

The paper is organized as follows. In Sect. 3.2, the MLS approximation and the FV-MLS method are briefly described. The new MLS-based sliding-mesh technique is presented in Sect. 3.3. Then, Sect. 3.4 is devoted to numerical simulations. Finally, the conclusions are drawn in Sect. 3.5.

3.2 MLS Approximations

The MLS approach was originally devised in 1981 for data processing and surface generation [7]. The basic idea of the MLS approach is to approximate a function $u(\mathbf{x})$ at a given point $\mathbf{x}$ through a weighted least squares fitting on a compact domain around $\mathbf{x}$. The resulting approximation, $\hat{u}(\mathbf{x})$, can be written as

$$u(\mathbf{x}) \approx \hat{u}(\mathbf{x}) = \mathbf{N}^T(\mathbf{x})\mathbf{u}_{\Omega_\mathbf{x}} = \mathbf{p}^T(\mathbf{x})\mathbf{M}^{-1}(\mathbf{x})\mathbf{P}_{\Omega_\mathbf{x}}\mathbf{W}(\mathbf{x})\mathbf{u}_{\Omega_\mathbf{x}} \tag{3.1}$$

where $\mathbf{N}^T(\mathbf{x})$ is the vector of MLS shape functions and $\mathbf{u}_{\Omega_\mathbf{x}}$ contains the known values of the function $u(\mathbf{x})$ at the compact domain $\Omega_\mathbf{x}$. These are the centroids of the cell and the centroids of the neighboring cells. These centroids form the stencil for a given cell. The size of the vector of the MLS shape functions is equal to the number of known values at the compact domain. The vector $\mathbf{p}^T(\mathbf{x})$ represents an m-dimensional functional basis (polynomial in this work), $\mathbf{P}_{\Omega_\mathbf{x}}$ is defined as a matrix where the basis functions are evaluated at each point of the stencil and the moment matrix $\mathbf{M}(\mathbf{x})$ is obtained through the minimization of an error functional [3, 9] and is defined as

$$\mathbf{M}(\mathbf{x}) = \mathbf{P}_{\Omega_\mathbf{x}} \mathbf{W}(\mathbf{x}) \mathbf{P}^T_{\Omega_\mathbf{x}} \tag{3.2}$$

The *kernel* or smoothing function, $W(\mathbf{x})$, plays an important role weighting the importance of the different points that take place in the approximation. A wide variety of *kernel* functions are described in the literature [8]. In this work the exponential *kernel* has been used, defined in one dimension as

$$W_j(x_j, x, s_x) = \frac{e^{-\left(\frac{d}{c}\right)^2} - e^{-\left(\frac{d_m}{c}\right)^2}}{1 - e^{-\left(\frac{d_m}{c}\right)^2}} \tag{3.3}$$

for $j = 1, \ldots, n_x$, where $d = |x_j - x|$, $d_m = 2\max\left(|x_j - x|\right)$.

In Eq. (3.3) d_m is the *smoothing length*, n_x is the number of neighbors and x is the reference point where the compact support is centered, the coefficient c is defined by $c = \frac{d_m}{s_x}$ and s_x is the shape parameter of the *kernel*. This parameter defines the kernel properties and therefore, the properties of the numerical scheme obtained.

As far as 2-D applications are concerned, a multidimensional *kernel* can be obtained by means of the product of 1D *kernels*. For instance, the 2D exponential *kernel* is defined by the following expression

$$W_j(x_j, y_j, x, y, s_x, s_y) = W_j(x_j, x, s_x) W_j(y_j, y, s_y) \tag{3.4}$$

The order of MLS approximations is determined by the polynomial basis used in the construction of MLS shape functions. Here we use quadratic and cubic polynomial basis. More details can be found in [1, 3, 17].

The high order approximate derivatives of the field variables $u(\mathbf{x})$ can be expressed in terms of the derivatives of the MLS shape function. So the n-th derivative is defined as

$$\frac{\partial^n \hat{u}}{\partial x^n} = \sum_{j=1}^{n_x} \frac{\partial^n N_j(\mathbf{x})}{\partial x^n} u_j \tag{3.5}$$

We refer the interested reader to [1, 3, 6] for a complete description of the computation of MLS derivatives.

3.3 Governing Equations and Numerical Method

In this section the numerical method will be presented. The integral form of two-dimensional Navier–Stokes equations over a domain of interest can be expressed in the Arbitrary Lagrangian–Eulerian (ALE) form as

$$\frac{\partial}{\partial t}\int_{\Omega}\mathbf{U}d\Omega+\int_{\Gamma}(\mathscr{F}(\mathbf{U}-\mathbf{V}^{\text{mesh}})-\mathscr{F}^{V}(\mathbf{U}))\cdot\mathbf{n}d\Gamma=\mathbf{0} \tag{3.6}$$

where $\mathbf{U}$ is the vector of conservative variables $\mathscr{F}=(\mathbf{F}_x,\mathbf{F}_y)$ is the inviscid flux vector, $\mathbf{V}^{\text{mesh}}=(u_{\text{mesh}},v_{\text{mesh}})^T$ is the mesh velocity, and $\mathscr{F}^V=\left(\mathbf{F}_x^V,\mathbf{F}_y^V\right)$ is the viscous flux. The vector $\mathbf{n}$ denotes the outer unitary normal to the interface Γ of the control volume Ω.

In order to develop a robust high order scheme for the resolution of the Navier–Stokes equations we have to distinct the different nature of the inviscid and viscous fluxes, as pointed in [3]. The viscous fluxes are evaluated directly at the integration points located at the interface Γ, but the inviscid fluxes are discretized using any suitable numerical flux. In order to solve the Riemann problem for the inviscid fluxes the variables at both sides of the interface need to be approximated.

For further details of the resolution procedure of the set of equations see [3, 6, 17].

3.3.1 *MLS-Based Sliding Mesh with Halo Cell*

The sliding mesh technique allows relative sliding motion of two grids. Here, we have focused on the case of one grid sliding over a static grid. The contact surface between both grids is called interface. Due to the relative movement between grids, the mesh is non-conformal. In order to deal with the non-conformality of the mesh at the interface we propose a new technique based on the use of MLS. A halo cell is used to reconstruct the variables needed to solve the Riemann problem. MLS approximations are used to compute the value of the variables at the centroid of the halo cell.

Let us consider a cell I. In order to solve the Riemann problem we require the value of the variables at both sides of the interface. As the mesh is non-conformal, this is not straightforward. To solve this, a halo cell is created in front of cell I. We call P_H to its centroid. The halo cell is a specular image of cell I, and the value of the variables at P_H is obtained from an MLS approximation as

$$\mathbf{U}_{P_H}=\frac{1}{A_{\text{halo}}}\int\mathbf{U}dA=\frac{1}{A_{\text{halo}}}\int\sum_{j=1}^{n_x}N_j(\mathbf{x}_{\text{halo}})U_j dA \tag{3.7}$$

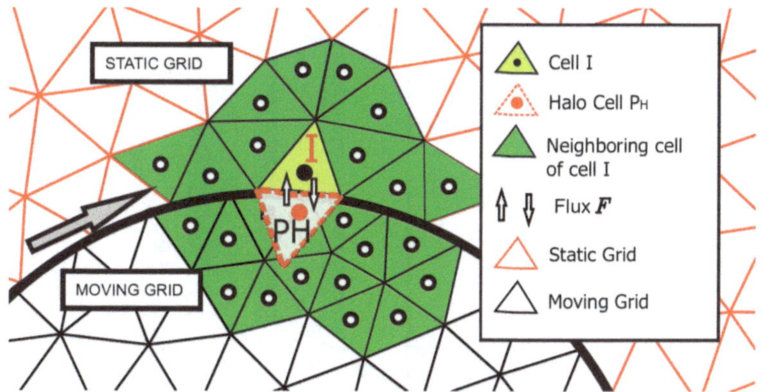

Fig. 3.1 Schematic representation of the interface Halo-Cell sliding mesh approach

For the MLS approximation at P_H we use the stencil of the cell I. Note that the computation of derivatives at P_H is straightforward using MLS. Once the value of the variables and the derivatives at the Halo cell centroid (P_H) is computed, it is used with the value of the centroid of cell I and its derivatives to solve the Riemann problem, as usual in a MUSCL framework.

The basic idea of this new approach is shown in Fig. 3.1.

The interface Halo-Cell sliding mesh approach simplifies the coding, specially in 3D, since it only requires the construction of a single halo cell, instead of the standard methodologies that require the computation of intersections between non edges (2D) or faces (3D) of the nonconforming elements at the interface.

Note that this method does not assure mass conservation. However, in the numerical tests we have observed that the magnitude of the conservation error is below the magnitude of the error in the variables, and the order of convergence is at least the same than for the variables. This is presented in the following section.

3.4 Numerical Results

This section presents numerical results for several test cases aimed at assessing the accuracy and efficiency of the proposed method for both steady and unsteady flow problems.

3.4.1 Ringleb Flow

The Ringleb flow test case [2] is used to evaluate the rate of convergence and the conservation error of the presented sliding mesh approach. The compressible

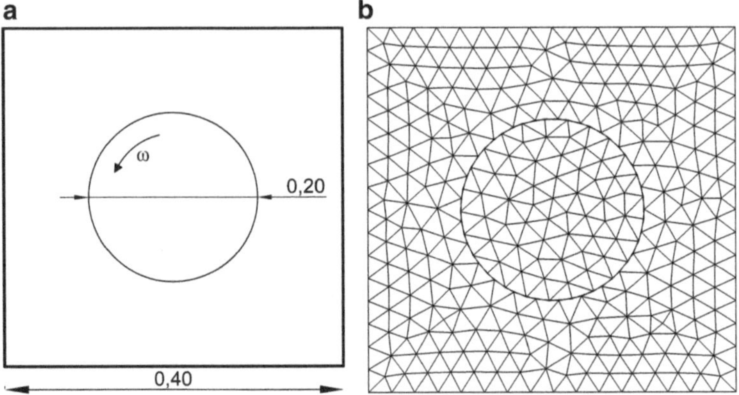

Fig. 3.2 (**a**) Basic geometry description of Ringleb flow problem. (**b**) Unstructured mesh of 580 triangles with non-conformal interface

Euler equations are solved on the computational domain $\Omega = [-1.15, -0.75] \times [0.15, 0.55]$. It is discretized using three different unstructured triangular meshes with 580, 2,270, and 9,044 control volumes. A schematic setup of the problem is shown in Fig. 3.2. In order to check the convergence order the L_2 norm of the entropy error is computed as

$$L_2^{\mathrm{ent}} = \sqrt{\frac{1}{\Omega} \int_\Omega \left(\frac{p/\rho^\gamma - p_\infty/\rho_\infty^\gamma}{p_\infty/\rho_\infty^\gamma} \right)^2 dx} \tag{3.8}$$

The convergence order of the L_2 norm of the entropy error and the conservation error are plotted in Fig. 3.3. Note that the expected order of convergence is recovered and notice that it is not affected by rotation. We also note that the magnitude of the conservation error is below the magnitude of the entropy error, and that it recovers the expected order.

3.4.2 One-Bladed Cross-Flow Turbine

In this section we present the results of the numerical method presented applied to the simulation of the unsteady viscous flow through a single bladed cross flow turbine. The basic configuration of this problem was defined by Ferrer in [4] and it is shown in Fig. 3.4. The turbine is formed with a single NACA 0015 airfoil with a chord $c = 1$ located at a radial distance $R = 2c$ from the center of rotation. The incompressible Navier–Stokes equations are solved using a higher-order method based on MLS with a collocated grid arrangement [17].

In this example, we present three different speed tip ratio ($\lambda = 1, 2, 5$), defined as $\lambda = \omega R / U_0$. The initial conditions are described in Table 3.1.

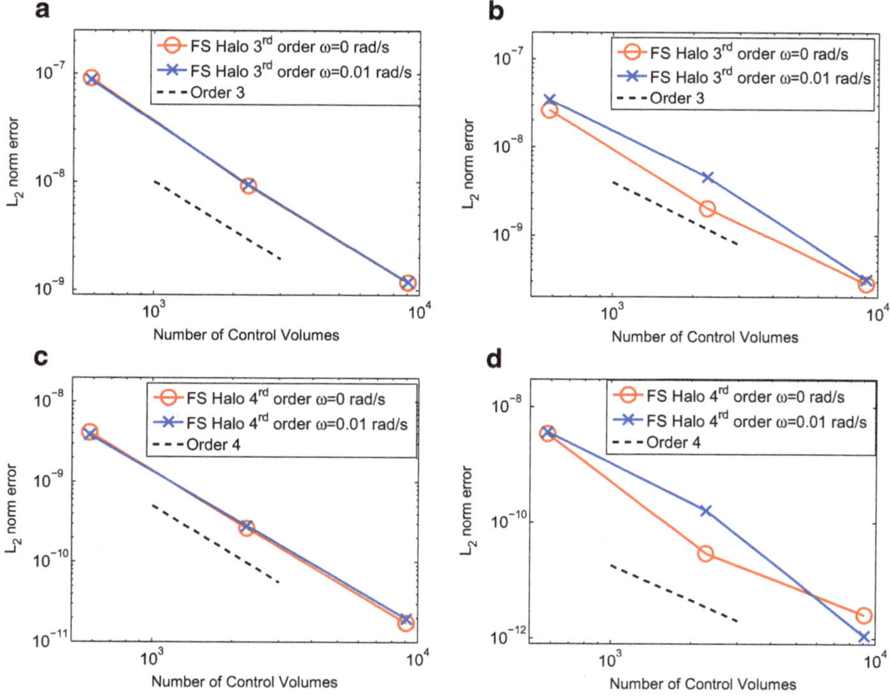

Fig. 3.3 L_2 norm of the entropy error (**a**, **c**) and conservation error (**b**, **d**) as function of the number of cells for the Ringleb flow for $\omega = 0.0$ rad/s and $\omega = 0.01$ rad/s for a third and a fourth order scheme

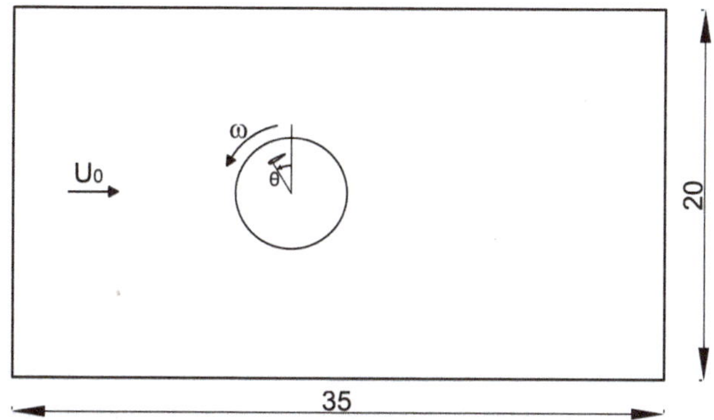

Fig. 3.4 Schematic description of the one-bladed cross-flow turbine

The force vector is computed in cartesian coordinates as

$$\mathbf{f} = \begin{Bmatrix} f_x \\ f_y \end{Bmatrix} = \oint (p\mathbf{n} - \nu(\nabla \mathbf{U} \cdot \mathbf{n}))d\Gamma \tag{3.9}$$

Table 3.1 Initial conditions for the one-bladed cross-flow turbine

Free-stream velocity U_0	Rotational speed ω	Tip speed ratio λ
0.2	0.5	5
0.5	0.5	2
1.0	0.5	1

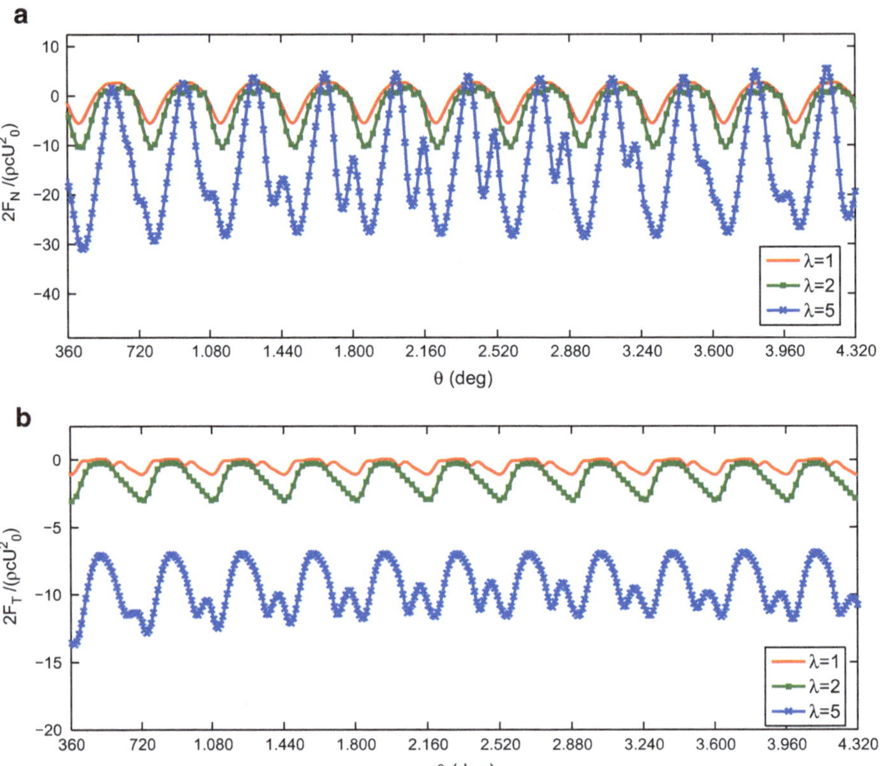

Fig. 3.5 (**a**) Normalized normal force and (**b**) normalized tangential force against azimuth for three different values of the tip speed ratio λ

where **n** is the outward pointing normal at the airfoil. Once the forces are obtained in Cartesian coordinates, they can be expressed on a normal-tangential frame of reference as

$$F_T = f_x \cos\theta + f_y \sin\theta \tag{3.10}$$

$$F_N = f_x \sin\theta - f_y \cos\theta \tag{3.11}$$

where θ is the angular location, as seen in Fig. 3.4.

Figure 3.5 compares the normalized tangential and normal forces, defined in Eq. (3.11), against the angular rotation θ.

Fig. 3.6 Velocity contours around the one-bladed cross-flow turbine. The solution is obtained with a 3rd FV-MLS method after six rotation cycles ($\theta = 1800°$) with $\lambda = 5$

The velocity contours near the NACA after six rotation cycles and $\lambda = 5$ are shown in Fig. 3.6. A smooth solution is obtained across the interface, and no numerical artifacts are observed. Our results present good agreement with the results obtained by Ferrer [4] using a 3rd order Discontinuous Galerkin method.

3.5 Conclusions

In this work a new high-order-preserving sliding-mesh methodology based on MLS approximations has been presented. The main advantage of this new approach is that it avoids the computation of intersections. Numerical results show that the proposed method keeps the order of convergence. In addition, the magnitude of the conservation error is below the magnitude of the L_2 norm in the variables. The new methodology has been applied to the computation of a single-bladed cross-flow turbine, and the results agree well with the results obtained with other sliding mesh approaches.

Acknowledgements This work has been partially supported by the *Ministerio de Ciencia e Innovación* (#DPI2010-16496), the *Ministerio de Economía y Competitividad* (#DPI2012-33622) of the Spanish Government, the *Consellería de Cultura, Educación e Ordenación Universitaria* of the *Xunta de Galicia* (grant # GRC2014/039) cofinanced with FEDER funds, the *Universidade da Coruña* and the grant *UDC-INDITEX*.

References

1. Chassaing JC, Khelladi S, Nogueira X (2013) Accuracy assessment of a high-order moving least squares finite volume method for compressible flows. Comput Fluids 71:41–53
2. Chiocchia G (1985) Exact solutions to transonic and supersonic flows. Technical Report AR-211, AGARD
3. Cueto-Felgueroso L, Colominas I, Nogueira X, Navarrina F, Casteleiro M (2007) Finite-volume solvers and moving least squares approximations for the compressible Navier–Stokes equations on unstructured grids. Comput Methods Appl Mech Eng 196:4712–4736
4. Ferrer E (2012) A high order discontinuous Galerkin–Fourier incompressible 3D Navier–Stokes solver with rotating sliding meshes for simulating cross-flow turbines. Ph.D., University of Oxford
5. Ferrer E, Willden RHJ (2012) A high order discontinuous Galerkin–Fourier incompressible 3D Navier–Stokes solver with rotating sliding meshes. J Comput Phys 231:7037–7056
6. Khelladi S, Nogueira X, Bakir F, Colominas I (2011) Toward a higher order unsteady finite volume solver based on reproducing kernel methods. Comput Methods Appl Mech Eng 200:2348–2362
7. Lancaster P, Salkauskas K (1981) Surfaces generated by moving least squares methods. Math Comput 87:141–158
8. Liu GR, Liu MB (2003) Smoothed particle hydrodynamics. A meshfree particle method. World Scientific Publishing, Singapore
9. Liu WK, Hao W, Chen Y, Jun S, Gosz J (1997) Multiresolution reproducing kernel particle methods. Comput Mech 20:295–309
10. Luo JY, Gosman AD, Issa RI, Middleton JC, Fitzgerald MK (1993) Full flow field computation of mixing in baffled stirred vessels. Chem Eng Res Design A 71(3):342–344
11. Nogueira X, Cueto-Felgueroso L, Colominas I, Khelladi S, Navarrina F, Casteleiro M (2010) Resolution of computational aeroacoustics problems on unstructured grids with a higher-order finite volume scheme. J Comput Appl Math 234(7):2089–2097
12. Nogueira X, Cueto-Felgueroso L, Colominas I, Gomez H, Navarrina F, Casteleiro M (2009) On the accuracy of finite volume and discontinuous galerkin discretizations for compressible flow on unstructured grids. Int J Numer Methods Eng 78:1553–1584
13. Nogueira X, Colominas I, Cueto-Felgueroso L, Khelladi S (2010) On the simulation of wave propagation with a higher-order finite volume scheme based on reproducing kernel methods. Comput Methods Appl Mech Eng 199:1471–1490
14. Rai M (1986) A conservative treatment of zonal boundaries for Euler equation calculations. J Comput Phys 62:472–503
15. Rai M (1986) A relaxation approach to patched-grid calculations with the Euler equations. J Comput Phys 66:99–131
16. Rai M (1987) Navier–Stokes simulations of rotor-stator interaction using patched and overlaid grids. J Comput Phys 3:387
17. Ramirez L, Nogueira X, Khelladi S, Chassaing JC, Colominas I (2014) A new higher-order finite volume method based on moving least squares for the resolution of the incompressible Navier–Stokes equations on unstructured grids. Comput Methods Appl Mech Eng 278: 883–901

Chapter 4
Vertical-Axis Wind Turbine Start-Up Modelled with a High-Order Numerical Solver

J.M. Rainbird, E. Ferrer, J. Peiro, and J.M.R. Graham

Abstract Vertical axis wind turbine (VAWT) start-up is a highly non-linear process, with turbines experiencing a long idling period of low acceleration before a sudden increase in rotational velocity to a final equilibrium state. The physics of start-up behaviour is not well understood, with some analyses showing VAWTs to be incapable of self-start, in spite of experimental and field evidence to the contrary. This study attempts to assess the impact of blade–wake interactions on start-up behaviour.

A high-order discontinuous Galerkin (DG) solver was used to simulate the flow around a VAWT, with experimentally obtained acceleration applied to the turbine, forcing it from standing to a tip-speed ratio (TSR) of 2 in a typically non-linear pattern. Additionally, the DG code was run with a constant TSR of 1, a TSR at which blade–wake interactions are known to occur. A blade-element momentum (BEM) method was used to provide an additional simulation of the turbine at this speed. Since the DG code captures blade–wake interactions while the BEM does not, a comparison of the outputs of the two was made to assess the impact of the interactions on blade forces.

4.1 Introduction

Vertical-axis wind turbines (VAWTs) rotate in a plane parallel to the flow, causing torques produced by blades to vary over the course of a rotation, with negative torque production in some sectors. The ability of fixed-blade Darrieus VAWTs to self-start is doubted by some, due to a predicted "dead band" of negative net torque production at certain low tip-speed ratios (TSRs) [1], though self-starting has been observed in controlled conditions [2, 3] and in field tests. The start-up phase of operation is

J.M. Rainbird (✉) • J. Peiro • J.M.R. Graham
Imperial College London, South Kensington Campus, London SW7 2AZ, UK
e-mail: j.rainbird11@imperial.ac.uk; j.peiro@imperial.ac.uk; m.graham@imperial.ac.uk

E. Ferrer
Universidad Politécnica de Madrid, Pza. de Cardenal Cisneros, 3, 28040 Madrid, Spain
e-mail: esteban.ferrer@upm.es

E. Ferrer, A. Montlaur (eds.), *CFD for Wind and Tidal Offshore Turbines*,
Springer Tracts in Mechanical Engineering, DOI 10.1007/978-3-319-16202-7_4

fundamental to design as moderate improvements to starting capability can improve a turbine's annual energy yields greatly [4] by allowing useful energy extraction in lower winds.

Little work has been done on the physics of VAWTs during start-up. Those studies that have successfully modelled the start-up behaviour of the Darrieus turbine in the time domain have relied on blade-element momentum (BEM) methods or similar [2, 3], with some also presenting CFD results at fixed, low TSRs [5, 6]. This study attempts to gain better understanding of the start-up process, in particular the blade/wake interactions that occur, by coupling together a CFD code and an experimentally obtained acceleration model to simulate start-up of a turbine from standing. Comparisons between the CFD code and a momentum method are also used.

The CFD code used in this work is a high-order discontinuous Galerkin (DG) solver for the Navier–Stokes equations, developed recently by Ferrer and Willden [7]. It is capable of working with sliding meshes, allowing high-order solutions of rotating bodies, and has been validated for a range of flows including Darrieus turbines [8]. The BEM code used is based on Paraschivoiu's double-multiple streamtube approach (DMS) [9]. Since the CFD solver provides a spatially discretised representation of the flow around the turbine and maintains wake structures in the flow, it can capture the impact of blade–wake interactions. The BEM code assumes a constant velocity across the rotor, with a calculated reduction in the free stream defined by an induction factor being the only consideration given to the wake, so it cannot capture these interactions. By comparing the output of the two codes, it is possible to highlight regions of the turbine's rotation that are particularly affected by blade–wake interactions.

4.2 Self-starting of Darrieus Turbines

In this study, self-starting ability is defined after Bianchini et al. [2] as when a turbine accelerates through its entire power curve to its fastest equilibrium state unaided. Figure 4.1 shows experimental results for TSR against time taken at Durham University [3, 10] at a selection of wind speeds; though in all cases the turbine rotates under its own power, it is only deemed to have self-started in flows with speeds above 6 m/s. The figure highlights the sensitivity of VAWTs to flow conditions: Rainbird's results demonstrate that a 20 % increase in flow velocity changes failure of a given turbine to self-start to a successful start-up.

The successful self-starting runs exhibit an idling period of slow acceleration until a TSR of approximately 1.6 has been reached when a rapid increase to the final running TSR occurs, while the unsuccessful run seems unable to overcome this idling phase. These marked differences in performance could be the result of blade Reynolds number effects, with blades only preforming well enough to

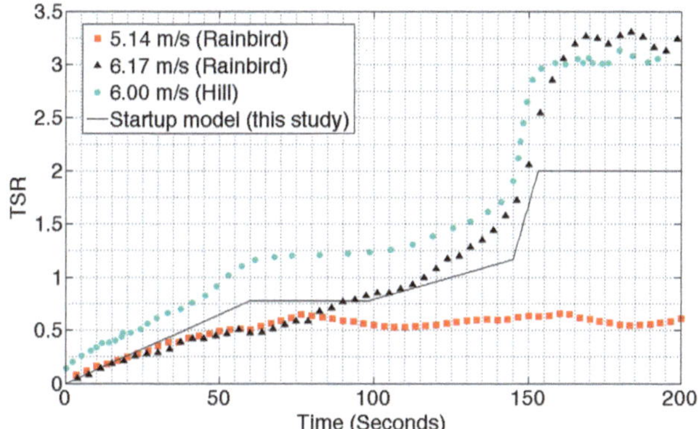

Fig. 4.1 Tip-speed ratio variation with time for a straight-bladed VAWT in flows of different velocities, after Rainbird [10] and Hill [3], alongside a model of Hill's data used in this study

achieve self-starting when a certain key relative flow velocity is reached. The sudden "jump" to higher rotational velocities could also be induced by beneficial blade–wake interactions at key values of TSR which this study sets out to investigate.

4.3 VAWT Aerodynamics

Due to the arrangement of the turbine's plane of rotation parallel to flow, the apparent velocity experienced by a VAWT blade, V_b, and its incidence angle, α, are dependent on its position during rotation, defined by the azimuth angle θ, see Fig. 4.2 for a blade velocity triangle. It can be seen that V_b and α are a function of rotational velocity ωR, plus the component of the flow through the turbine, V_W, resolved into a blade chord-wise direction using the azimuth, together with the blade-normal component of V_W. Note that by comparison, HAWT V_b and α are defined only by ωR and V_W.

This dependence on azimuth means that blade effective angle of incidence is varying, even at a constant wind speed and TSR. At higher TSRs where the rotational velocity component comes to dominate the velocity triangle there is less incidence variation, but during start-up a wide range of incidences is present. Figure 4.3 shows lift, drag and tangential coefficients against incidence for the NACA 0015, a profile commonly used in VAWTs, taken from [11]. The incidences experienced by VAWT blades at different TSRs have been highlighted: at a TSR of 4, it can be seen that these do not extend much past stall points, but at TSR = 1, angles as high at $\pm 90°$ are encountered. At this TSR, the tangential coefficient (which provides turbine torque) looks to be close to zero if integrated around a full rotation of the turbine, hence the low acceleration at TSRs around 1, demonstrated in Fig. 4.1.

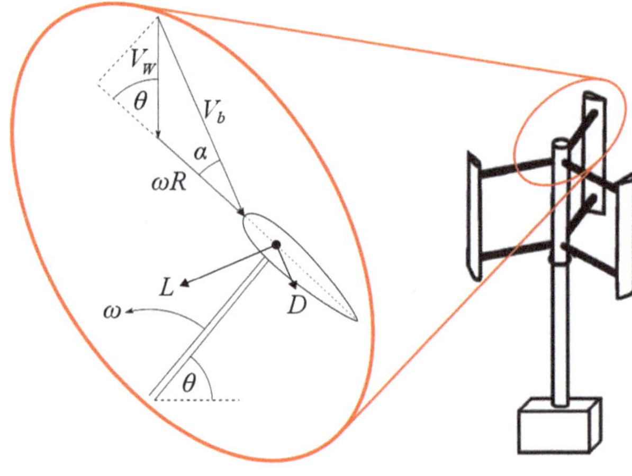

Fig. 4.2 VAWT velocity triangle

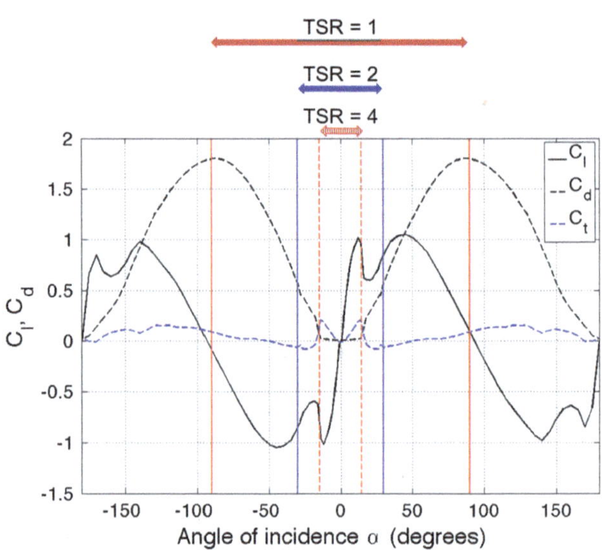

Fig. 4.3 Lift, drag and tangential coefficients against incidence [11], with limits in VAWT blade incidence at different TSRs highlighted

4.4 Approaches to the Problem

Much existing VAWT research is concerned with optimising peak performance, focussing on high TSR ranges [12–14]. Generally values of TSR < 2 can be considered relevant to turbine start-up, depending somewhat on blade chord to turbine radius ratio c/R, and may be called low TSR. In the studies that do cover these relevant TSRs, reliance is usually placed in either BEM methods or CFD.

4.4.1 Existing Studies

Most CFD studies consider fixed TSRs. Plots of blade forces with azimuth are commonly given [5, 6, 15], with Rosetti and Pavesi going further and including some flow visualisation for the case at TSR = 1, in the form of vorticity distributions and shear stress vector plots [5].

To the best of the authors' knowledge, the only numerical methods that have yielded time-dependent analysis of VAWTs to date are BEM methods [2, 5] or simpler methods neglecting induced velocities [3]. These codes can struggle to accurately model the self-starting ability of the turbines due to a "dead band" of negative net torques that can be demonstrated for VAWTs using a geometric analysis where induction factors are ignored [1]. At low TSRs, blade loadings are low and so induction factors are low, meaning the outputs of BEM codes are close to this simplified analysis. This "dead band" involves torques which are certainly low, as shown in Fig. 4.3, and is the cause of the idling period seen in Fig. 4.1, but the fact that turbines are capable of self-start in some conditions means that this integrated torque is not always negative.

This limitation at low TSRs impacts on the usefulness of BEM design tools in the start-up phase of operation. The BEM studies that have successfully modelled turbine start-up have required modification of blade characteristics in order to match experiments: Rossetti and Pavesi modify their aerofoil coefficient data and introduce rapid fluctuations in the free stream of the order of 5 and 10 % in order to "start" their model [5], while Bianchini et al. replace the original blades with conformally transformed, cambered equivalents above a specified TSR [2] after Migliore's "virtual camber" theory [16].

These two referenced BEM studies use developments of Paraschivoiu's double-multiple streamtube (DMS) approach, which accounts for a decrease in flow velocities across the rotor by modelling the upstream and downstream halves of the turbine as separate actuators. However, the codes cannot capture blade/wake interactions and so cannot provide any insight into the impacts of these on the start-up process. They are robust and computationally cheap, hence their popularity for analysis of VAWTs and HAWTs.

CFD could give some indication of the significance of these interactions, with Rossetti and Pavesi's study demonstrating that there is some impact at TSR = 1 [5]. CFD simulations are computationally costly however, hence the paucity of studies attempting to replicate anything more than a few TSRs.

4.4.2 Approach of the Current Study

In developing the DG solver used in this study, a single-bladed and a three-bladed VAWT were analysed at TSRs of 1, 2 and 5 [8]. Differences between the forces generated by a single blade of the turbine in the single-blade and three-blade configurations were shown. Assuming that blade-blade interference is negligible, these differences must be due to interactions between the measured blade and the additional wakes of the other blades in the three-bladed case, suggesting the importance of blade–wake interactions.

This study extends on earlier work [8] by running the DG solver with variable TSRs in an attempt to qualitatively assess the impact of blade/wake interactions on the start-up characteristics of VAWTs, with particular focus on the non-linearity of the start-up ramp shown in Fig. 4.1 and the inability to self-start in low free stream velocities.

The DG code has been adapted to model start-up by forcing its acceleration to follow a time history based on experimental data. Since the acceleration is forced, this paper does not investigate inertia effects.

An investigation has also been undertaken into turbine performance at TSR = 1 by comparing the output of the DG code and a BEM code, both for constant TSR. This TSR was chosen as it has been identified as one at which blade–wake interactions occur [8] and is within the “dead band” at which low torques are produced during start-up. This comparison gives an indication of the impact of blade–wake interactions since the DG code captures them while the BEM does not.

4.4.3 BEM Method Code

A code based on Paraschivoiu’s DMS BEM model [9] is used in this study. Figure 4.4 shows a diagram of a VAWT and its representation in the DMS code. The flow is discretised into a number of streamtubes (shown separated by dashed lines), allowing the model to capture stream velocity variations across the turbine but not including wake expansion effects. The VAWT is modelled as two separate actuators

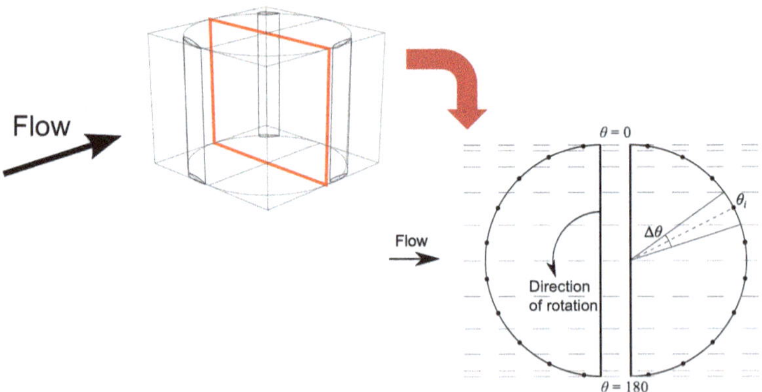

Fig. 4.4 Double-multiple streamtube configuration: on the *left* is a sketch of a VAWT, on the *right* is its representation in a 2-D DMS BEM code

in an array, one for the upstream pass of the blades through the flow and one for the downstream pass, effectively splitting the analysis of the two halves of the turbine. The exit velocity of an upstream streamtube is equal to the inlet velocity of the streamtube directly downstream from it, giving an equilibrium velocity between the two actuators. This requires an assumption that there is sufficient distance between the actuators for the equilibrium velocity to develop fully.

The two halves of the analysis are shown in Fig. 4.4, with the point at which the split lies in relation to the sketched turbine marked on the sketch. The flow is assumed to be two-dimensional and blade end effects are ignored. Forces are averaged over each streamtube, with increased accuracy the thinner the streamtube.

4.4.4 Discontinuous-Galerkin Solver

The DG solver is described briefly below, for full details on it see [7, 8]. The solver is high order and is capable of handling a rotating portion of mesh within the main static mesh. This is achieved using curved element boundaries and hanging nodes.

A second-order stiffly stable finite difference method is used for temporal discretisation of the Navier–Stokes equations. For spatial discretisation in the x–y plane, the symmetric interior penalty Galerkin formulation with modal basis functions is used, allowing the hanging nodes and sliding meshes necessary for the rotating portion of the mesh without the need for mortar type techniques. Spatial discretisation in the z direction is provided by spectral methods, but as for the BEM analysis, only two dimensional results are presented here and end effects are ignored.

Blade forces are calculated by integrating pressures around the blade. The method maintains wake structures well, making it suitable for the study of blade/wake interactions. Turbines modelled in this study are single bladed, so any wake encountered in the downstream pass of the blade is that of its own upstream pass.

4.4.5 Experimental Data Used for Modelled Acceleration

There is very little time-dependent data in existence for VAWTs covering the start-up period. Figure 4.1 contains results of two studies completed using the same apparatus, namely Durham University's 2m wind tunnel and a straight-bladed giromill with three NACA-0018 profile blades. The turbine has a diameter of 760 mm and blades have 80 mm chords and 600 mm spans. In both studies, the turbine was allowed to rotate freely from standing with no load attached. Also included in Fig. 4.1 is the acceleration model used in this study, based on Hill's data [3].

4.4.6 Equivalence of Codes in the Study at TSR = 1

For the comparative study of the output of the BEM and DG codes, care has been taken over equivalence of code inputs. For turbine dimensions and blade number this is straightforward enough, both were set to model the same single bladed VAWT with matching turbine radius of 2 units, blade chord of 1 unit and NACA 0015 profile. Ensuring blade performance is similar in both codes is more challenging. In the DG code, blade shape is defined as a boundary in the mesh and the code does the rest in terms of calculating flows around it and forces generated by it. As mentioned, for the BEM code these forces are calculated from input coefficients, usually obtained experimentally. Here, they have instead been obtained by using the DG code to model an isolated NACA 0015 aerofoil. This is mounted at the quarter chord in the centre of the rotating mesh which is held stationary, as opposed to the turbine model where the aerofoil is mounted offset from the centre by the turbine radius and the mesh rotates to set the required TSR. A set of steady aerofoil lift and drag curves were computed in the range $0°$ to $+90°$, since the range of incidences experienced at a TSR of 1 is $-90°$ to $+90°$ and the profile is symmetrical.

The blade Reynolds number of the flow used in the DG code when modelling the full turbine is 100. Currently, code runs have been limited to low Reynolds numbers to reduce the expense of computations. These are unrealistic values for even a small domestic VAWT. For the Durham University giro mill used to obtain the experimental data used in this study Reynolds numbers would be around three orders of magnitude higher, but qualitatively useful outputs have been obtained. Though the blade forces calculated at these low Reynolds numbers are not similar to those produced in real VAWTs, blades still shed vorticity into the free-stream as the blade lift forces change with time, allowing blade–wake interactions to be studied. An added benefit of this low Reynolds number limitation is that it removes the uncertainties of turbulence modelling.

In the course of the turbine's rotation at TSR $=$ 1, the blade's relative velocity will be around twice that of the free stream at an azimuth of $0°$, and roughly zero at an azimuth of $180°$. To account for these variations, the NACA 0015 force coefficients were taken for the required range of incidences at Reynolds numbers of 50, 100, 150 and 200. For the runs at Reynolds numbers higher than 50, vortex shedding was induced at high incidences. The oscillations in blade forces caused by the vortex shedding were smoothed out before use in the BEM code.

The BEM code calculates the Reynolds number of the flow apparent to the blade and the blade incidence, with a bilinear interpolation used to provide approximate blade lift and drag from the NACA 0015 forces obtained from the DG code.

The NACA 0015 experiences rectilinear flow when modelled on its own. When the full VAWT is modelled, the flow apparent to the blade is curvilinear, following the circumference of the rotation. Migliore proposed that this imparts a virtual camber and incidence onto any profile used in a VAWT [16], with a recent experimental and numerical study supporting the proposal [17]. In order to correct for the use of the rectilinear flow NACA 0015 in the BEM code, where the VAWT blade should also experience a curved flow, corrections based on those of Migliore have been applied as follows.

The VAWT "virtual blade" profile is calculated by conformally transforming the NACA 0015 so that its camber line follows an arc of its circumference of rotation. The quarter chord of the virtual profile is coincident with the unmodified profile, but leading and trailing edges curve away from it, the trailing edge curving away more, as it is further from the quarter chord than the leading edge. Virtual incidence is calculated by taking the angle between the chord line of this modified profile and the original aerofoil's chord line. Here, with a chord of 1 and radius of 2, the virtual incidence is 7.16°. The virtual camber is represented by a lift coefficient shifted such that at zero incidence, $C_l = 0.2$. This value has been estimated from experimental data taken for a NACA 0018 and a conformally transformed equivalent [17]. Work is ongoing on DG code modelling of the conformally transformed profile to improve on this crude camber adjustment.

4.5 Results

4.5.1 Modelled Turbine Acceleration

Figure 4.5 shows results of the study where the DG code simulation of the turbine had Hill's experimentally obtained acceleration [3] applied to it, see also Fig. 4.1. Note that the blade force coefficients are normalised using the free-stream rather than the blade relative velocity. There is some evidence of blade–wake interactions taking place in the force plots, the slight dips at the positive peaks in normal force can be indicative of this [8]. Blade–wake interactions were confirmed at these points by examining pressure contours from the code. These dips are present throughout

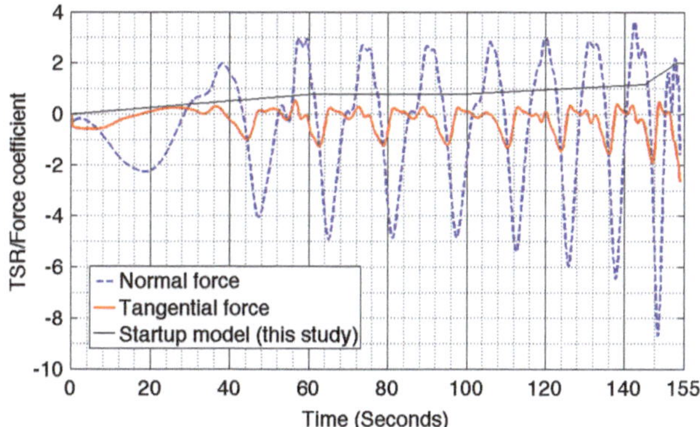

Fig. 4.5 Normal and tangential blade coefficients vs time for the DG code modelled VAWT with acceleration per start-up model (shown here, see also Fig. 4.1)

the idling period and the acceleration that follows it, suggesting that blade–wake interactions occur throughout the start-up process beyond the initial ramp up to idling.

It should be noted that the experimental data used to produce the acceleration model was obtained using a turbine with dimensions and blade profile different to that of the DG code turbine simulation. The impact of the different dimensions has been accounted for by reducing the magnitude of the acceleration in the model compared to the original experimental data (the DG code turbine has a larger c/R ratio, so it would be expected to reach an equilibrium state at a lower TSR). Similar sized turbines using NACA 0015 and NACA 0018 blade profiles exhibit similar start-up characteristics [18], suggesting that the use of the model is valid in spite of the different blade profiles used. In any case the start-up acceleration here has a much longer time-scale than the blade cycling and hence the exact acceleration is probably unimportant. The experiment also used a three-bladed turbine, which would obviously be expected to experience more blade–wake interactions than the one-bladed DG code turbine. However the blade–wake interactions captured in the model would be present in the real turbine when any of the blades are interacting with their own wakes.

4.5.2 BEM Method and CFD Solver Comparison at TSR = 1

Figure 4.6 shows results of the comparative study of the BEM and DG codes over the course of one rotation at a steady TSR $= 1$. Previous work with the DG code has shown that blades are expected to interact with their own wakes at TSRs between 0.5 and 1.6, and that these interactions should primarily affect the region of azimuths between 180° and 270° [8]. The results presented here clearly support these findings, with reasonable agreement between the codes in the range 0–170°. The codes differ in both force coefficient values and plot shape in the region where blade–wake interactions are expected, the DG code capturing a trough and a peak in the force

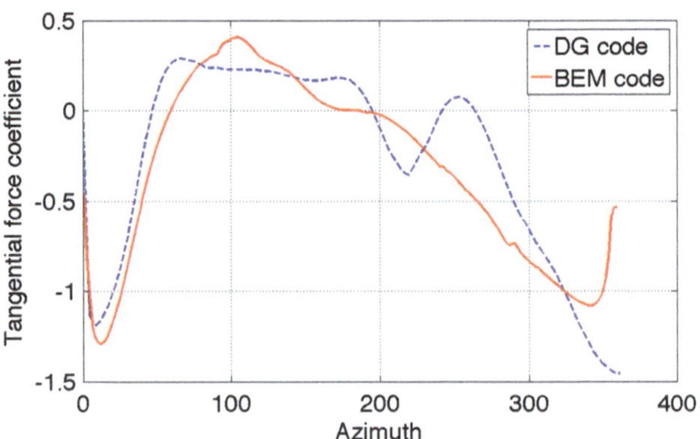

Fig. 4.6 Tangential force coefficients vs azimuth over a single rotation from the DG and BEM codes

while the BEM output is steadier. Efforts have been made to ensure the BEM code accounts for variation in blade incidence and relative velocity, and flow curvature as described above; given this and the results of existing research, these differences can be confidently ascribed to the impact of blade–wake interactions. Beyond 270°, the agreement between the codes is not strong but plot shapes are similar, both exhibiting a steady downward trend to a negative peak.

Differences here are due to the DMS method used in the BEM code. At azimuths near to zero, approximately 0–30° and 330–360°, flow velocities relative to the blade are high, so the forces induced on it are high. This gives high induction factors, meaning much of the energy in the wind is extracted in the upstream half of the analysis before it reaches the downstream half (note that while the magnitude of the BEM tangential force is smaller than the force of the DG code between 330° and 360°, it is larger between 0° and 30°).

4.6 Conclusions

Computational studies have been performed on the start-up of a VAWT in an attempt to assess the impact of blade–wake interactions on the highly non-linear acceleration behaviour of the turbines. A high order DG solver has been used to provide simulations of a turbine accelerating from a standing start to a TSR of 3. An experimentally obtained acceleration model was applied to the simulation of the turbine, with outputs of blade tangential and normal forces produced. These force plots show evidence of blade–wake interactions for much of the start-up in the form of characteristic perturbations, the presence of these interactions being confirmed by examining pressure contour plots of the turbine where these perturbations occur. They all also occur in the range of azimuths ($180° < \theta < 270°$) at which blade–wake interactions are expected.

An additional study was performed at a constant TSR of 1, a TSR within the range experienced during the non-linear start-up behaviour, and one at which previous studies have identified the occurrence of blade–wake interactions. The DG solver was again used to simulate VAWT performance, with a BEM code used alongside it. The DG code is capable of simulating blade–wake interactions while the BEM code cannot. The outputs of both codes have been compared, with good agreement between the outputs in the upstream half of the turbine's rotation (at azimuths of at 0–180°), where no blade–wake interactions are expected. Blade–wake interactions should occur at azimuths of 180–270° and the code outputs are very different here. This suggests that the method of comparing BEM and CFD outputs could be developed into a means of quantifying the impact of blade–wake interactions on VAWT blade performance.

References

1. Watson G (1979) The self-starting capabilities of low-solidity fixed pitch Darrieus rotors. In: Wind energy workshop, 1st, Cranfield, Beds
2. Bianchini A, Ferrari L, Magnani S (2011) Start-up behavior of a three-bladed H-Darrieus VAWT: experimental and numerical analysis. In: Proceedings of the ASME Turbo Expo, Vancouver
3. Hill N, Dominy R, Ingram G, Dominy J (2009) Darrieus turbines: the physics of self-starting. Proc Inst Mech Eng Part A 223(A1):21–29
4. Worasinchai S, Ingram G, Dominy R (2011) The effects of improved starting capability on energy yield for small HAWTs. In: Proceedings of the ASME Turbo Expo, Vancouver
5. Rossetti A, Pavesi G (2013) Comparison of different numerical approaches to the study of the H-Darrieus turbines start-up. Renew Energ 50:7–19
6. Beri H, Yao Y (2011) Double multiple stream tube model and numerical analysis of vertical axis wind turbine. Energ Power Eng 3(3):262–270
7. Ferrer E, Willden RHJ (2012) A high-order discontinuous Galerkin Fourier incompressible 3D Navier-Stokes solver with rotating sliding meshes. J Comput Phys 231(21):7037–7056
8. Ferrer E (2012) A high-order discontinuous Galerkin Fourier incompressible 3D Navier-Stokes solver with rotating sliding meshes for simulating cross-flow turbines. Ph.D. thesis, Oxford University
9. Paraschivoiu I (1981) Double-multiple streamtube model for Darrieus wind turbines. In: Proceedings of 2nd DOE/NASA wind turbines dynamics workshop, Cleveland
10. Rainbird J (2007) The aerodynamic development of a vertical axis wind turbine. MEng thesis, University of Durham
11. Sheldahl RE, Klimas PC (1981) Aerodynamic characteristics of seven symmetrical airfoil sections through 180-degree angle of attack for use in aerodynamic analysis of vertical axis wind turbines. Sandia National Laboratories SAND80-2114
12. Strickland J, Smith T, Sun K (1981) Vortex model of the Darrieus turbine: an analytical and experimental study. Sandia National Laboratories SAND81-7017
13. Oler J, Strickland JH, Im BJ, Graham GH (1983) Dynamic stall regulation of the Darrieus turbine. Sandia National Laboratories SAND83-7029
14. Danao LA, Qin N, Howell R (2012) A numerical study of blade thickness and camber effects on vertical axis wind turbines. Proc Inst Mech Eng Part A 226(7):867–881
15. Gupta R, Biswas A (2010) Computational fluid dynamics analysis of a twisted three-bladed H-Darrieus rotor. J Renew Sustain Energ 2:043111
16. Migliore PG, Wolfe WP, Fanucci JB (1980) Flow curvature effects on Darrieus turbine blade aerodynamics. J Energ 4(2):49–55
17. Bianchini A, Balduzzi F, Rainbird J, Peiro J, Graham JMR, Ferrara G, Ferrari L (2015) An experimental and numerical assessment of airfoil polars for use in Darrieus wind turbines. Part 1—flow curvature effects. ASME Journal of Turbomachinery (in press)
18. Worasinchai S, Ingram GL, Dominy RG (2014) The physics of H-Darrieus turbine starting behaviour. In: Proceedings of the ASME Turbo Expo 2014, Dusseldorf

Chapter 5
Large-Eddy Simulation of a Vertical Axis Tidal Turbine Using an Immersed Boundary Method

Pablo Ouro Barba, Thorsten Stoesser, and Richard McSherry

Abstract Vertical Axis Tidal Turbines (VATTs) are an innovative way of harnessing renewable energy from tidal streams. Herein a novel numerical approach using a refined Large Eddy Simulation (LES) code to simulate the performance of a VATT is presented. The turbine blades are modelled with Lagrangian markers using the Immersed Boundary Method which offers several advantages especially concerning computational effort. Comparisons of the LES results with experimental and numerical data suggest reasonably good accuracy of the code. In addition, the stability of the method for high Reynolds number flows is also discussed.

5.1 Introduction

In recent years the pressing need to find alternative sources of renewable energy has given impetus to research in new fields. The fact that tidal energy is more predictable than, for example, wind energy makes it a promising renewable energy source. Vertical Axis Tidal Turbines (VATTs) are an attractive way to harness the hydrokinetic energy from tidal streams, however the technology is considered in its infancy and much research is still needed to improve current designs.

Experimental testing of VATTs cannot provide a full explanation as to why certain turbine design parameters such as blade shape, angle of attack or number of blades work better than others. In addition, the complexity of the flow through and around the turbine cannot be predicted with analytical tools and hence requires more detailed predictions. Thus, high-resolution numerical simulations using Computational Fluid Dynamics (CFD) can provide an enormous amount of data and, if accurate, can give insight into the complex fluid–structure interaction of tidal

P. Ouro Barba (✉) • T. Stoesser • R. McSherry
Hydro-Environmental Research Group, School of Engineering, Cardiff University, Cardiff, UK
e-mail: OuroBarbaP@cardiff.ac.uk; Stoesser@cardiff.ac.uk; McSherryR@cardiff.ac.uk

E. Ferrer, A. Montlaur (eds.), *CFD for Wind and Tidal Offshore Turbines*,
Springer Tracts in Mechanical Engineering, DOI 10.1007/978-3-319-16202-7_5

turbines. Moreover, CFD simulations might be cheaper and less time-consuming than experiments and hence can provide quick answers to design variations.

There has been some work on numerical simulation of VATTs, most CFD studies to date have employed Reynolds Averaged Navier–Stokes (RANS) based CFD models [9, 10] and only very few have used LES [4, 8]. The disadvantage of RANS models is that the flow is temporally averaged and thus unsteady vortices shed off turbine blades are damped and their influence reduced. Large-Eddy Simulations achieve a more accurate representation of the instantaneous flow [14, 18], which appears to be critical for VATT simulations, because the downstream side of a VATT is affected significantly by the vortices shed on the upstream side.

In all above-mentioned studies, the sliding mesh technique is used to represent the movement of the blades in relation to the fixed Eulerian fluid mesh. This technique requires small time steps to fulfil the numerical requirements. In this paper, a VATT is simulated using the Immersed Boundary (IB) method which can be considered a Eulerian–Lagrangian method. The fluid flow is solved on a fixed (Eulerian) Cartesian grid and the turbine movement is accomplished by Lagrangian meshes which conform to the shape of the blades of the VATT. This reduces the complexity of LES simulations because re-meshing of the Eulerian grid is not required. On the other hand, there are some weaknesses regarding the accuracy of the Eulerian–Lagrangian method compared to body-fitted sliding mesh methods: the boundary layer of the blade is not properly resolved and there are artificial residual velocities inside the Lagrangian domains. A refined 2D version of the LES code HYDRO3D is used in this research. HYDRO3D has previously been well validated in a number of other studies of complex flows [1, 6, 7, 17, 19]. The code's Immersed Boundary Method implementation has also been thoroughly validated [5]. Although turbulence is not perfectly represented in a 2D LES, it still appears superior to 2D RANS based methods because of its inherent abilities to represent the dominating instantaneous horizontal vortices of this flow. The objectives of this study are to assess the accuracy of the LES code for such flows and to investigate numerical aspects with which code performance is enhanced.

5.2 Numerical Model

The formulation for the Eulerian and Lagrangian parts is presented first. The code HYDRO3D is based on finite differences with staggered storage of the Cartesian velocity components on Cartesian grids. Second-order central differences are used to approximate convective and diffusive terms and the fractional-step method is used with a three-step Runge–Kutta predictor. The solution of a Poisson pressure-correction equation is achieved using a multi-grid method in the final step as a corrector of predicted velocities. The WALE subgrid scale model [11] is used to approximate the subgrid scale stresses. The numerical scheme for the Runge–Kutta step k reads:

$$\frac{\tilde{u}^k - u^{k-1}}{\Delta t} = \alpha_k C u^{k-1} + \beta_k C u^{k-1} + \frac{1}{2}\alpha_k D(u^{k-1} + \tilde{u}^k) - \alpha_k \frac{\partial p^{k-1}}{\partial x} \tag{5.1}$$

$$\nabla^2 \phi^k = \frac{\nabla(\tilde{u}^k + \Delta t \cdot f^k)}{2\alpha_k \Delta t} \tag{5.2}$$

$$u^k = \tilde{u}^k - \alpha_k \Delta t \frac{\partial \phi^{k-1}}{\partial x} \tag{5.3}$$

$$p^k = p^{k-1} + \phi^k - \frac{\alpha \Delta t}{2Re} \nabla^2 \phi^k \tag{5.4}$$

where $\tilde{u}^k$ is the predicted Eulerian velocity, u^{k-1} is the Eulerian velocity at the previous step, ϕ is the pseudo-pressure, p is the pressure and f^k is the force due to the IB. C and D represent the convective and diffusive operations. The Runge–Kutta coefficients are $\alpha_1 = \beta_1 = 1/3, \alpha_2 = \beta_2 = 1/2$ and $\alpha_3 = \beta_3 = 1$.

5.2.1 Immersed Boundary

The Immersed Boundary [12] method is based on the direct forcing approach introduced by Fadlun et al. [2] and improved by Uhlmann [20]. It is based on the transmission of the Eulerian velocities to the Lagrangian markers. The reaction force f is calculated and then added as a source/sink in Eq. (5.2). Equations (5.5)–(5.7) summarize the process to compute the value of the reaction force.

$$U_L^k = \sum_{j=1}^{N_E} \tilde{u}^k \cdot \delta(x_j - X_L) \cdot (\Delta x \cdot \Delta y) \tag{5.5}$$

$$F_L^k = \frac{U_L^* - U_L^k}{\Delta t} \tag{5.6}$$

$$f^k(x_j) = \sum_{L=1}^{N_L} F_L^k \cdot \delta(X_L - x_j) \cdot \Delta V_L \tag{5.7}$$

where x_j, $\tilde{u}$ and f represent position, velocity and force in the Eulerian field, and X_L, U and F do the same for the Lagrangian domain. The Eulerian cell size is represented by $\Delta x \cdot \Delta y$ and the area assigned to each Lagrangian particle is

ΔV_L. The delta function δ represents the interpolation function and is presented in Eq. (5.8). The kernel chosen is the ϕ_4^* derived by Yang et al. [22]. This kernel is smoother and has a wider support than other kernels [13, 15] which is translated in a smoother transition of the interpolations and reduction of spurious force oscillations.

$$\delta(x_j, y_j) = \frac{1}{\Delta x \cdot \Delta y} \cdot \phi_4^* \left(\frac{x_j - X_L}{\Delta x} \right) \cdot \phi_4^* \left(\frac{y_j - Y_L}{\Delta y} \right) \tag{5.8}$$

5.2.2 *Parallelization*

The code presents an efficient hybrid Message Passing Interface (MPI) and OpenMP parallelization, Fraga et al. [3]. The fluid flow is solved in a rectangular Cartesian domain which is parallelized using MPI, and split into several blocks assigned to different processors. Although communication between blocks is required, this type of parallelization notably reduces the computational effort for the pressure Poisson equation solver using multi-grid. OpenMP is used for the Lagrangian Immersed Boundary part. The IB loops with large numbers of particles are divided into chunks and assigned to different threads. When a large number of IB points is used, the Lagrangian part becomes the computationally restrictive part of the simulation, e.g. the time required on the IB is higher than that required for the pressure solver. Thus, a proper hybrid parallelization is needed for an optimum performance. For further details of this parallelization strategy, the interested reader is referred to [3].

5.2.3 *VATT Physics with IB*

Immersed boundaries are used to represent the blades of the VATT. The simulated turbine follows a prescribed counter-clockwise circular movement at constant rotational velocity. The position (X_L, Y_L) and velocity (U_L^*, V_L^*) of each Lagrangian marker at any time step is determined from Eqs. (5.9) and (5.10); the velocity is then fed into Eq. (5.6) in order to compute the force exerted by the Lagrangian point.

$$\begin{pmatrix} X_L \\ Y_L \end{pmatrix} = \begin{pmatrix} -R_L \cdot \sin(\alpha_{0_L} + \alpha) \\ R_L \cdot \cos(\alpha_{0_L} + \alpha) \end{pmatrix} \tag{5.9}$$

$$\begin{pmatrix} U_L^* \\ V_L^* \end{pmatrix} = \begin{pmatrix} -R_L \cdot \Omega \cdot \cos(\alpha_{0_L} + \alpha) \\ -R_L \cdot \Omega \cdot \sin(\alpha_{0_L} + \alpha) \end{pmatrix} \tag{5.10}$$

where R_L and α_{0_L} are the radius and initial angle of the Lagrangian marker L, α is the angle rotated by the turbine, and Ω is the rotational speed.

The forces F_x and F_y obtained in Eq. (5.6) are decomposed in order to obtain corresponding drag and lift forces on the blades. These are then used to compute the torque at the shaft and the power coefficient (Cp) [Eq. (5.11)] is obtained.

$$C_P = \frac{N \cdot T \cdot \Omega}{R} \tag{5.11}$$

where T is the torque, R is the turbine radius and N is the number of blades.

5.3 Results

Large-Eddy Simulations obtained using the present methodology are compared with experimental and numerical data from Maitre et al. [9] and McNaughton et al. [10]. The comparison is focused on power coefficient variation as a function of tip speed ratio. Both comparative studies employed RANS based methods. Maitre et al. used the $k - \omega$ SST turbulence model while McNaughton et al. used the $k - \omega$ SST LRN model. Appreciable differences between the predictions of these models and those obtained using LES are observed: 2D LES is shown to offer fairly good agreement with the experimental data.

In the second part of this study, special attention is paid to numerical parameters that affect the IB stability. Also, the influence of the time integration on the CFL condition is studied. In order to ensure the no-slip condition on the IB, multi-direct forcing (MDF) [21] is applied using different numbers of loops.

5.3.1 Model Validation

In [9] a vertical axis turbine with three NACA 0018 blades is tested experimentally and numerically. Here, the VATT is scaled to a chord length of 1 m (0.032 m in [9]). Simulation details are presented in Table 5.1. The numerical domain has a length of 32 m in the x-direction and 18 m in the y-direction. The Eulerian cell size of 0.01 m is constant throughout the domain, giving a total of 5.76 million cells. Each of the blades is discretized by 1,478 IB points (Fig. 5.1).

The power coefficient from the simulated cases with the five TSRs outlined in Table 5.1 is shown in Fig. 5.2. LES results show fairly good agreement with the

Table 5.1 Simulation setup details

Inlet velocity	2.3 m/s
Reynolds number	73600
Tip speed ratios	1, 1.5, 2, 2.5, 3
Chord length	1 m
Angle of attack	0°
Turbine radius	2.73 m
Solidity	0.175

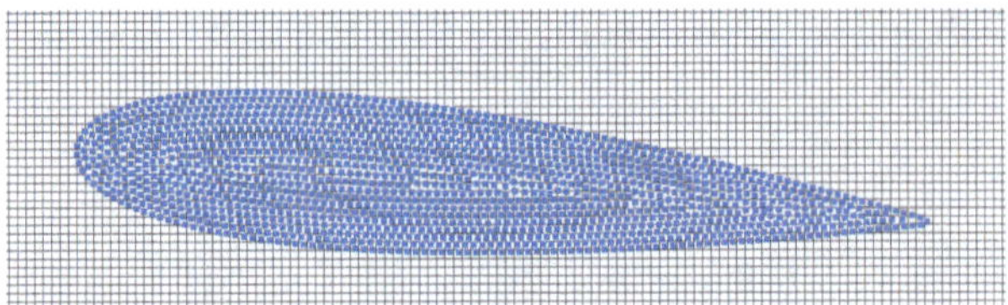

Fig. 5.1 NACA 0018 conformed by Lagrangian points representing the IB

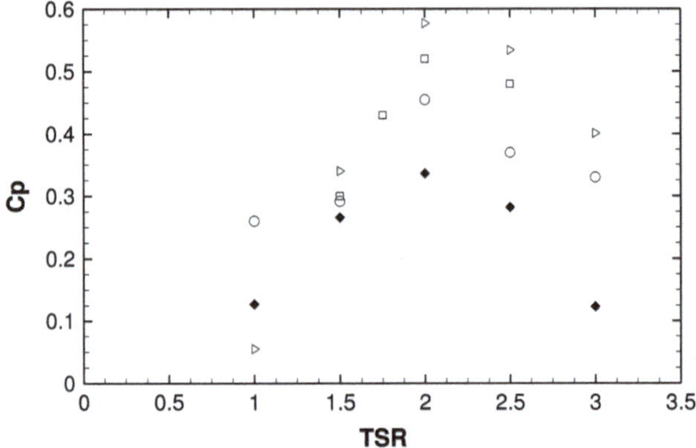

Fig. 5.2 Performance curve. *Filled diamond*: experimental [9], *right triangles*: $k - \omega$ SST [9], *squares*: $k - \omega$ SST LRN [10], *circle*: present 2D LES results

experimental results, with improvement from the RANS results for all TSRs except $\lambda = 1$. On the other hand, the RANS simulations notably overestimate the Cp value for TSRs larger than 2. Although the more turbulent RANS approach offered by the $k-\omega$ SST LRN noticeably lowers the Cp value, the difference with the experimental results is still significant.

Figure 5.3 shows the total Cp of the turbine as a function of the rotated angle. The LES results fit quite nicely the experimental results from [9]. A possible explanation of the reduction of the mean Cp value compared to other 2D simulations is that the IB method exhibits minimal porosity of the blade, i.e. some fluid enters the blade, thereby slightly reducing the resulting drag.

5.3.2 Time Integration

In the present simulation model the Lagrangian grid and the Eulerian field are decoupled. The turbine rotation is implicitly determined at each new time step as rotational speed is fixed. This can cause the CFL condition to be violated by any of the IB points. If any of the Lagrangian markers has a large displacement, this can cause serious force oscillations. In order to analyse this, predictions of Cp using different time integrations are presented here.

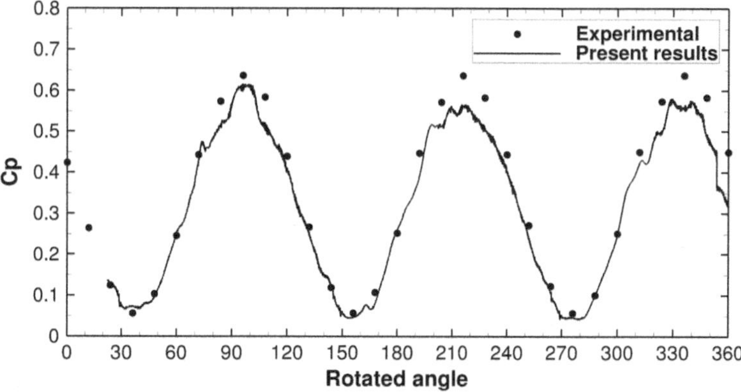

Fig. 5.3 Power coefficient curve for the three blades. Experimental [9] and present results

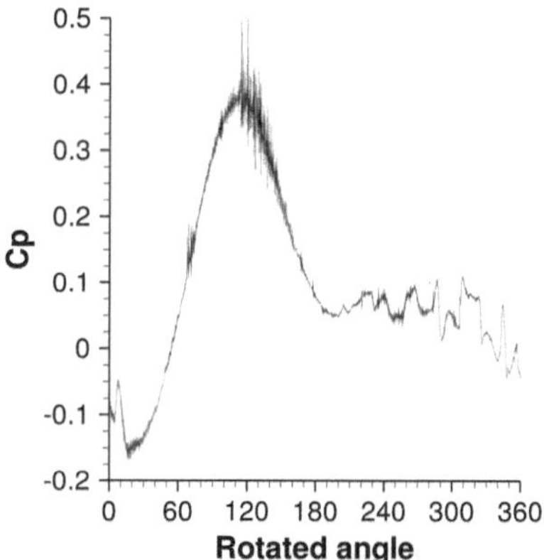

Fig. 5.4 CFL = 1.00

Figures 5.4, 5.5 and 5.6 show the stability dependence of the force oscillations regarding the CFL condition, as already stated by Shin et al. [16]. This is due to the fact that the Eulerian field and the Lagrangian markers are decoupled and the CFL condition can be violated in terms of the Lagrangian points. It is readily seen that a CFL value of 0.75–0.85 guarantees a good stability of the IB procedure with respect to the computation of forces. On the other hand, a CFL equal to 1.0 produces large force oscillations and less stability on the resulting forces.

5.3.3 *Multi-Direct Forcing Sensitivity*

Multi-Direct Forcing [21] is used to enforce the no-slip condition on the Lagrangian markers when the direct forcing method is used. Figure 5.7 presents the magnitude

Fig. 5.5 CFL = 0.85

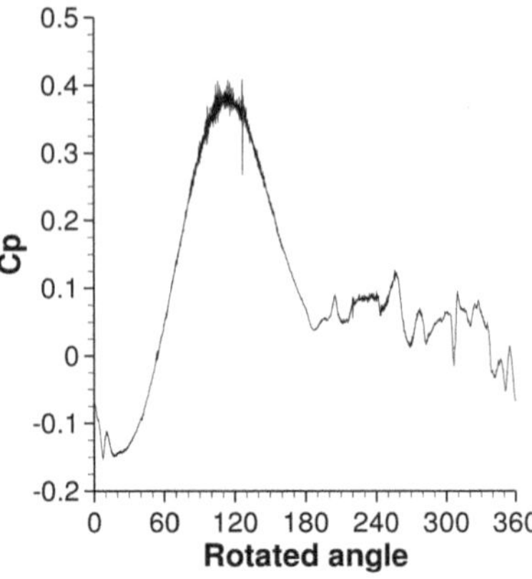

Fig. 5.6 CFL = 0.75

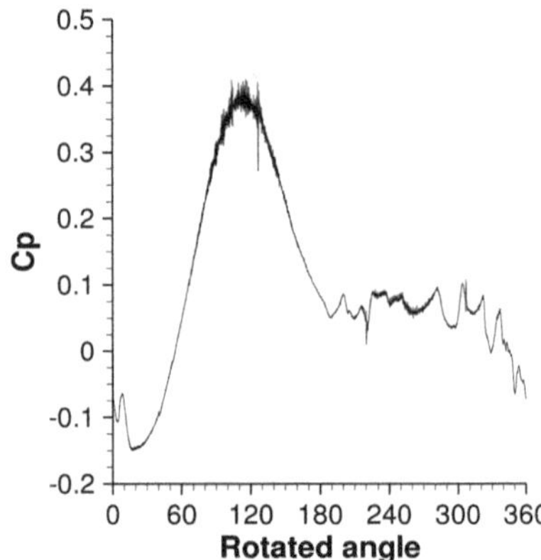

Fig. 5.7 l_2-norm using multi-direct forcing

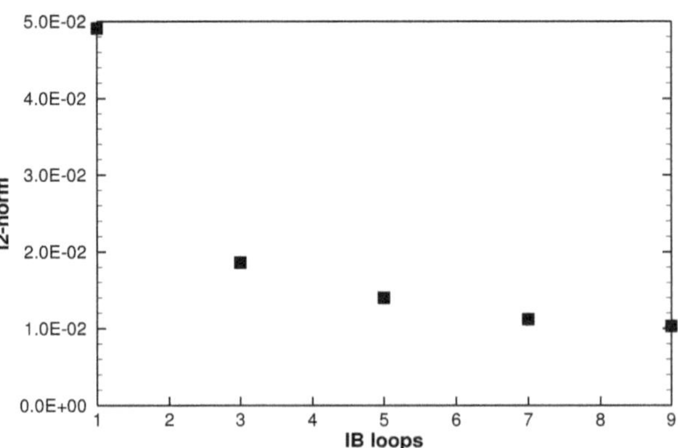

of the residual velocity [Eq. (5.12)] within one of the blades of the turbine for five different states of the MDF. The more loops done within the IB procedure the better the no-slip condition is stipulated. However, this leads to a larger computational expense and a slowdown of the calculations [3]. A large improvement is achieved when three MDF loops are used in comparison with the simple direct forcing approach (one loop). Note that afterwards the difference is reduced with an approximately linear convergence. Thus there is not a large difference when using five loops rather than nine in terms of stability

$$l_2\text{–norm} = \sqrt{\frac{\sum_{L=1}^{N}[(U_L^k - U_L^*)^2 + (V_L^k - V_L^*)^2]}{N}}. \tag{5.12}$$

5.4 Conclusions

2D Large-Eddy Simulation of the flow past a vertical axis turbine using the Immersed Boundary method has been shown to give reasonably good agreement with experimental data. Compared with RANS-based simulation studies found in the literature, a closer match to the experimental data in terms of power coefficient magnitude and behaviour has been achieved. Thus, the presented model is considered to predict reasonably well the drag and lift forces on the blades in addition to resolving quite accurately the shedding of vortices off the blades, their advection with the residual flow and their influence on the downstream section of the turbine.

However, it is demonstrated that the presented IB method requires advanced numerical treatment in order to reduce the force oscillation. The relevance of the time integration and multi-direct forcing is demonstrated; these are important ingredients to ensure accurate predictions of forces acting on the blades.

Acknowledgements The authors wish to thank Cardiff University's ARCCA and HPC Wales for providing High Performance Computing resources.

References

1. Bomminayuni S, Stoesser T (2011) Turbulence statistics in an open-channel flow over a rough bed. ASCE J Hydraul Eng 137:1347–1358
2. Fadlun EA, Verzicco R, Orlandi P, Mohd-Yusof J (2000) Combined immersed boundary finite difference methods for three dimensional complex flow simulations. J Comput Phys 161:35–60
3. Fraga B, Ouro P, Stoesser T (in revision) Speedup of an Eulerian-Lagrangian large-eddy simulation solver by hybrid MPI-OpenMP parallelization. Comput Fluids
4. Iida A, Kato K, Mizuno A (2007) Numerical simulation of unsteady flow and aero-dynamic performance of vertical axis wind turbines with LES. In: 16th Austral-Asian fluid mechanics conference, Crown Plaza, Gold Coast
5. Kara MC, Stoesser T (2014) A strong FSI coupling scheme to investigate the onset of resonance of cylinders in tandem arrangement. In: ASME 2014 33rd International conference on ocean, offshore and arctic engineering, San Francisco, 8–13 June 2014, Paper No. OMAE2014-23972
6. Kara S, Stoesser T, Sturm TW (2012) Turbulence statistics in compound channels with deep and shallow overbank flows. J Hydraul Res 50:482–493

7. Kim D, Kim JH, Stoesser T (2013) The effect of baffle spacing on hydrodynamics and solute transport in serpentine contact tanks. J Hydraul Res 51:558–568
8. Li C, Zhu S, Xu Y, Xiao Y (2013) 2.5D large eddy simulation of vertical axis wind turbine in consideration of high angle of attack flow. Renew Energy 51:317–330
9. Maitre T, Amet E, Pellone C (2013) Modeling of the flow in a Darrieus water turbine: wall grid refinement analysis and comparison with experiments. Renew Energy 57:497–512
10. McNaughton J, Billard F, Revell A (2014) Turbulence modelling of low Reynolds number flow effects around a vertical axis turbine a a range of tip-speed ratios. J Fluids Struct 47:124–139
11. Nicoud F, Ducros F (1999) Subgrid-scale stress modelling based on the square of the velocity gradient tensor. Flow Turbul Combust 62:183–200
12. Peskin CS (1972) Flow patterns around heart valves: a digital computer method for solving the equations of motion, Ph.D. thesis, Albert Einstein College of Medicine
13. Peskin CS (2002) The immersed boundary method. Acta Numer 11:479–517
14. Rodi W, Constantinescu G, Stoesser T (2013) Large-Eddy simulation in hydraulics. IAHR monographs. CRC Press, Boca Raton
15. Roma A, Peskin CS, Berger J (1999) An adaptive version of the immersed boundary method. J Comput Phys 153:509–534
16. Shin SJ, Huang W, Sung HJ (2008) Assessment of regularized delta functions and feedback forcing schemes for an immersed boundary method. Int J Numer Methods Fluids 58:263–286
17. Stoesser T (2010) Physically realistic roughness closure scheme to simulate turbulent channel flow over rough beds within the framework of LES. ASCE J Hydraul Eng 136:812–819
18. Stoesser T (2014) Large-eddy simulation in hydraulics: Quo Vadis? J Hydraul Res 52:441–452
19. Stoesser T, Nikora V (2008) Flow structure over square bars at intermediate submergence: Large Eddy Simulation (LES) study of bar spacing effect. Acta Geophys 56:876–893
20. Uhlmann M (2005) An immersed boundary method with direct forcing for the simulation of particles flow. J Comput Phys 209:448–476
21. Wang Z, Fan J, Luo K (2008) Combined multi-direct forcing and immersed boundary method for simulating flows with moving particles. Int J Multiphase Flow 34:283–302
22. Yang X, Zhang X, Li Z, He G (2009) A smoothing technique for discrete delta function with application to immersed boundary method in moving boundary simulations. J Comput Phys 228:7821–7836

Chapter 6
Computational Study of the Interaction Between Hydrodynamics and Rigid Body Dynamics of a Darrieus Type H Turbine

Omar D. Lopez, Diana P. Meneses, and Santiago Lain

Abstract The present study discusses two-dimensional numerical simulations of a cross-flow vertical-axis marine (Water) turbine (straight-bladed Darrieus type) with particular emphasis on the turbine unsteady behavior. Numerical investigations of a model turbine were undertaken using commercial computational solvers. The domain and mesh were generated using a glyph script in POINTWISE-GRIDGEN, while the simulations were performed in ANSYS-FLUENT v14. For the simulation, a sliding mesh technique was used in order to model the rotation of the turbine; a shear stress transport k–ω turbulence model was used to model the turbulent flow. In order to simulate the interaction between the dynamics of the flow and the Rigid Body Dynamics (RBD) of the turbine a User Define Function (UDF) was generated. The primary turbine operational variables of interest were the evolution of torque, power, and runaway speed. Numerical results show that as the freestream velocity is increased, the runaway angular speed of the turbine increases, which is consistent with the observation that the frequency of oscillation of the angular velocity (in the quasi steady-state) increases as the freestream velocity also increases. For a given turbine, it was observed that the increment in the moment of inertia of the turbine does not influence the average value of the runaway angular velocity (quasi-steady state) but causes an increase in the time taken for achieving this quasi steady-state.

O.D. Lopez (✉)
Universidad de los Andes, Cra 1 Este N° 19A-40, Bogota, Colombia
e-mail: od.lopez20@uniandes.edu.co

D.P. Meneses
Universidad de San Buenaventura, Cra 8H N° 172-20, Bogota, Colombia
e-mail: dmeneses@usbbog.edu.co

S. Lain
Universidad Autonoma de Occidente, Calle 25 N° 115-85, Cali, Colombia
e-mail: slain@uao.edu.co

E. Ferrer, A. Montlaur (eds.), *CFD for Wind and Tidal Offshore Turbines*,
Springer Tracts in Mechanical Engineering, DOI 10.1007/978-3-319-16202-7_6

6.1 Introduction

The scarcity of no-renewable energy sources and the awareness of environmental problems such as global warming has intensified the search for new and better energy sources, which must be more environmentally friendly. The kinetic energy of sea water contained in marine currents is one of the key elements in this quest with a very high potential of success. Turbines used in marine applications can be classified into two types: Horizontal axial turbine, which has the axis of rotation in the direction of flow (e.g., Tidal turbines), and vertical axis turbine which has the axis of rotation perpendicular to the water flow. The latter is ideal when the current changes in direction as they do not need a yaw mechanism.

The study of flow dynamics in vertical turbines has had a great interest in recent years by the scientific community, particularly in relation to improving efficiencies and understanding of their performance. In the specific case of the Darrieus type turbine, it still requires a deeply understanding of its performance especially for hydrodynamic applications. Computational Fluid Dynamics (CFD) has proven to be a useful tool for flow analysis around Darrieus type turbines. For example, regarding performance prediction of Darrieus turbines, Lain et al. [1] show numerical results for the moment coefficient (C_m) using CFD that are in satisfactory agreement with experimental data. Ferreira [2] also shows and discusses an experimental–computational study on the aerodynamics and performance of a small scale vertical axis turbine obtaining two important conclusions: Experimentally, it was observed that the roughness of the surface finish blades has significant effects on the turbine performance, and computationally it was observed that the predictions found in the 3D model are significantly lower than in 2D by the presence of tip vortices. Regarding Rigid Body Dynamics (RBD) coupling with CFD in the simulation of vertical axis wind turbines, Lee [3] implemented a computational model of a Savonius turbine including an equation for the resistive torque for direct-drive generator showing good agreement with theory and experiments. The implemented model was able to not only estimate the transient output power but also predict whether the turbine was able to rotate or not under certain initial condition. Untaroiu et al. [4] computationally studied the self-starting capabilities of a vertical axis wind turbine using the commercial software ANSYS-CFX [5], in order to do so the ANSYS programming language CEL was used to couple the RBD of the turbine with CFD. Numerical results show an accurate prediction of the operating speed with an error of less than 12 % with respect to experimental data. Finally, Belhache et al. [6] performed a computational and numerical study of a loaded Darrieus marine current turbine. One of the main purposes of this study was to determine a correction factor f needed to overcome the differences between the predicted torque of the 2D model and the actual torque of the three-dimensional experimental setup.

Self-starting capabilities of vertical axis water turbine have not been highly studied and as it is well known one of the disadvantages of the Darrieus turbine is precisely its poor capabilities for self-starting. Given the need that still exists in the scientific community by the application of CFD models in the study of vertical turbines, especially Darrieus type, the present work aims to study the interaction between the hydrodynamics and the RBD of a Darrieus type H turbine using a two-dimensional CFD model.

6.2 Numerical Simulation Methodology

6.2.1 Domain and Mesh

The present work uses as base case the turbine model described in Lain et al. [7], which has been previously studied both computationally and experimentally [1, 8]. The geometrical and operational parameters of this base case are specified in Table 6.1.

The computational domain used in Lain et al. [7] and also in the present study for the simulation of the turbine is shown in Fig. 6.1. The domain is subdivided into two: one domain near the turbine which is rotational and the other far from the turbine which is fixed. Between the two domains exists an interface that allows continuous flux of mass and momentum. The rotational domain has the following dimensions $D_e = 2.8\ R$ and $D_i = 1.1\ R$, while the fixed domain has $h = 10\ R$ and $l = 16\ R$.

Table 6.1 Geometrical and operational parameters of base case

Parameter	Value
Radius (R)	0.45 m
Number of blades	3
Airfoil profile	NACA0025
Chord length	132.75 mm
Tip speed ratio (λ)	1.745
Angular velocity (ω)	6.28 rad/s
Solidity	0.89
Wingspan	0.7 m
Reference area (A)	0.63 m^2
Fluid	Water @ 20 °C

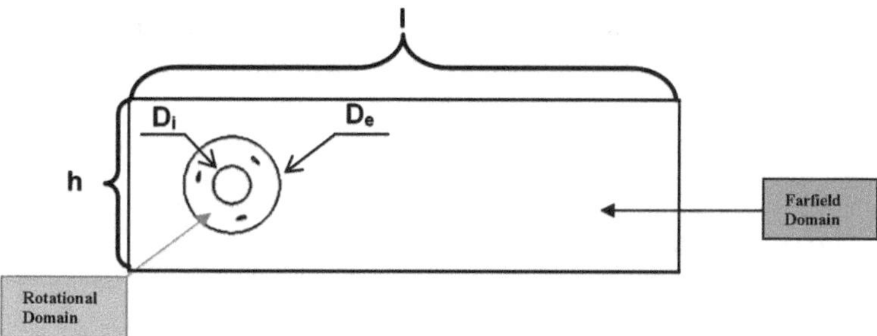

Fig. 6.1 Computational domain

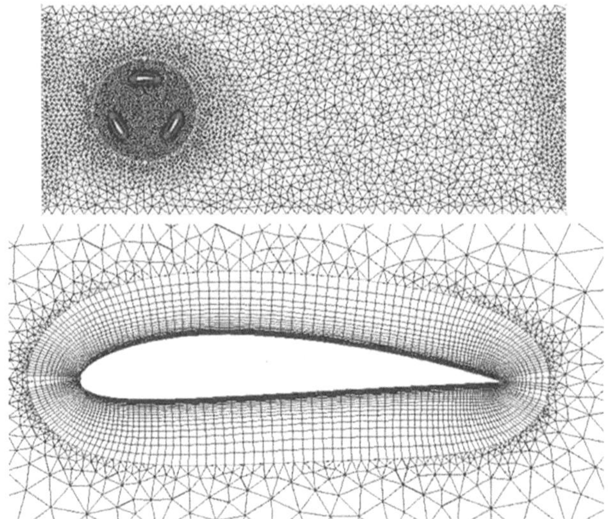

Fig. 6.2 Mesh details

Once the size of the computational domain was defined a mesh was generated using the commercial software POINTWISE-GRIDGEN [9]. The mesh generation starts with the upper and lower surfaces of the airfoil which are divided into a fixed number of elements. From these surfaces and using a hyperbolic mesh generation scheme a structured grid is created close to the airfoil surface. This approach allows the control of the size of the first element and the growth rate of the mesh in the normal direction to the surface which is important in order to correctly capture the boundary layer. The rest of the rotational domain is meshed with triangular elements (unstructured grid) in which the growth rate between the boundary layer region and the rotating domain is also controlled in order to achieve a smooth transition between the structured and unstructured meshes. Finally, the outer domain was meshed also with triangles and ensuring a smooth transition at the interface. Figure 6.2 shows some details of the generated mesh in which it is observed its characteristics close to the airfoil surface. Through a convergence analysis, it was concluded that a mesh of about 100k elements was sufficient to ensure that the results are independent of the mesh size at a moderate computational cost (see Lain et al. [7] for more details).

6.2.2 Solver Setup

The different boundary conditions for the computational domain were set as follows: Velocity inlet at the left edge of domain with a given freestream velocity (V_∞); Moving walls were specified at the top and bottom planes of the domain with a translational speed equal to V_∞; Pressure outlet was specified at the right edge of the domain, with a value equal to the atmospheric pressure; Finally Walls (non-slip) were specified at the surfaces of the turbine blades. The commercial software ANSYS-FLUENT v14 [5] was used as a flow solver. A scheme of sliding mesh between fixed and rotating domains was used in order to simulate the rotation of the blades. An incompressible, Newtonian, and turbulent flow was assumed and the turbulence model selected for the simulations was k–ω SST, the choice is based on previous research work (see Lain et al. [1, 7]). Spatial and temporal discretizations of second order for all equations and the SIMPLE (Semi-Implicit Method for Pressure Linked Equations) scheme for the solution of the governing equations were used. To advance in time, a time step (Δt) of 0.005 s was used and 40 iterations were performed per time step. The size of the time step was also obtained from a convergence analysis as shown in Lain et al. [1].

Two output variables are of special interest to the present study: Hydrodynamic torque (T_h) generated and the runaway angular velocity of the turbine. The hydrodynamic torque will be presented in non-dimensional form (moment coefficient C_m) as is typically found in the literature. The moment coefficient is given by Eq. (6.1)

$$C_m = \frac{T_h}{0.5 \cdot \rho \cdot V_\infty^2 \cdot R \cdot A} \tag{6.1}$$

where ρ is the density of the fluid and T_h is the hydrodynamic torque.

The RBD of the turbine is given by an ordinary differential equation (see Eq. (6.2) where I is the moment of inertia of the turbine) which has to be coupled with the flow solver. Since no model for the load (generator) was implemented then $\sum T = T_h$.

$$\sum T = I \frac{d\omega}{dt} \tag{6.2}$$

The RBD-CFD coupling was implemented in a loose way using a User Defined Function (UDF), which is programmed in C. UDFs are defined using macros called DEFINE which are predesigned in FLUENT for different actions to be executed. Two macros were used for the RDB-CFD coupling `DEFINE_CG_MOTION` and `DEFINE_EXECUTE_AT_END`.

`DEFINE_EXECUTE_AT_END` is executed at the end of each time step and it is used to calculate the instantaneous hydrodynamic torque (T_h) of the turbine and to numerically solve Eq. (6.2). The numerical solution of the RBD equation is obtained with an explicit Euler scheme. `DEFINE_CG_MOTION` macro is used to update the angular velocity and to control the instantaneous motion of the rotating domain.

6.3 Results

Initially, the implemented model was validated with the base case and the results were compared with other numerical and experimental results. Details of the computational results of the validation are shown in Lain et al. [7] in which a good agreement was observed. Once the model was validated, the implementation of the UDF was tested for some cases in which two numerical parameters (freestream velocity and turbine moment of inertia) were changed in order to observe their influence in the turbine performance specially in the runaway angular velocity of the turbine.

6.3.1 Runaway Angular Velocity for the Base Case

The first test performed in the implemented model was the estimation of the runaway velocity of the turbine of the base case. It was assumed that the blades were made of polyester resin so that the moment of inertia of the turbine relative to its center of rotation was 1.3 kg/m^2. The initial condition of the flow was impulsively started and at the same time the turbine started to rotate from rest. Figure 6.3 shows the evolution of the hydrodynamic torque, it is clear that during the first second there is a transient evolution of the turbine torque until it reaches a periodic behavior (quasi steady-state). The averaged output moment is zero because there is not a resistive torque (runaway condition) but the instantaneous maximum torque coefficient is approximately 0.1.

Figure 6.4 shows the evolution of the angular velocity of the turbine. It is observed that the angular velocity starts at rest (initial condition) and that the angular velocity increases very rapidly, so that in less than 0.5 s it reaches a maximum angular speed of 19 rad/s. An overshoot in the evolution of the runaway angular velocity of the turbine is appreciated. Between 0.5 and 1 s the angular velocity decreases in a transient evolution until it reaches a periodic behavior. The average runaway angular velocity of the turbine is approximately 14.3 rad/s.

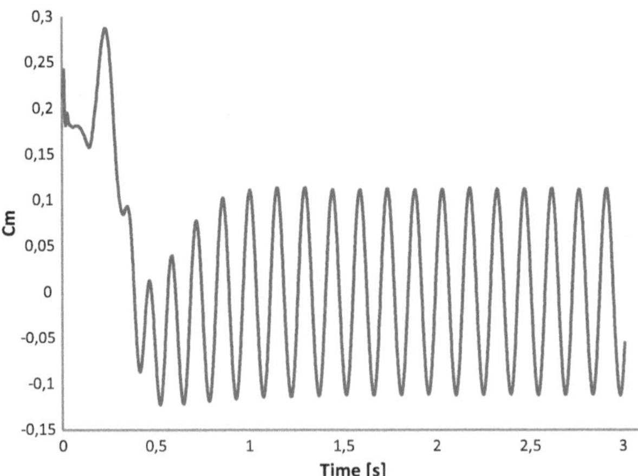

Fig. 6.3 C_m evolution for the base case

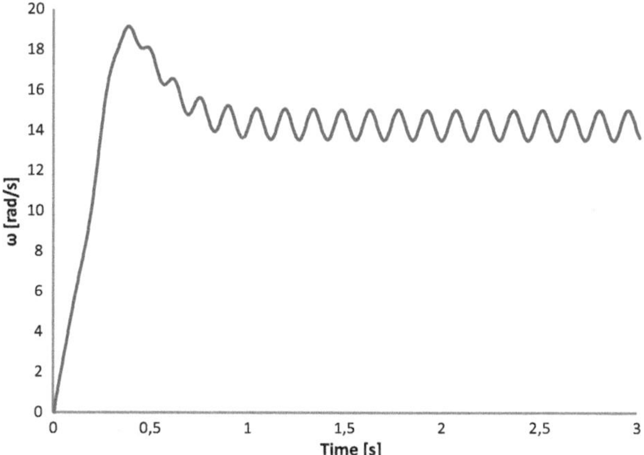

Fig. 6.4 Angular velocity evolution for the base case

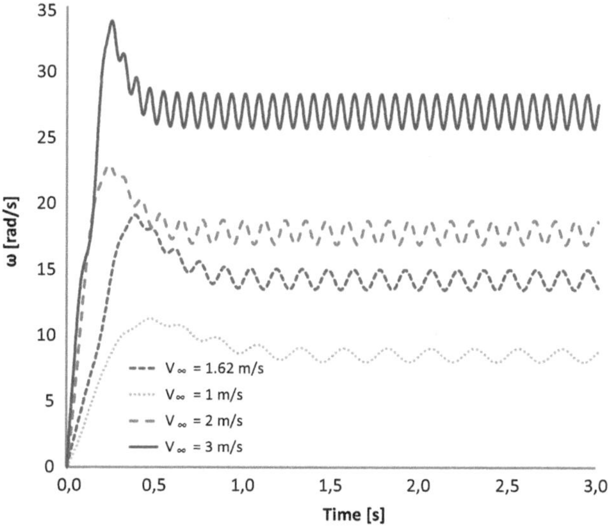

Fig. 6.5 Angular velocity evolution for different freestream velocities

6.3.2 Influence of the Freestream Velocity in the Runaway Angular Velocity

The base case turbine was tested with four different freestream velocities (1, 1.62, 2, and 3 m/s) and the self-starting evolution was analyzed. Figure 6.5 shows that the average runaway angular velocity increases as the free stream velocity increases. It is clear that the frequency of oscillation of the evolution of the angular velocity increases as the freestream velocity also increases, this was also observed in the evolution of the moment coefficient (not shown). A Fourier analysis of these curves shows that the frequencies of oscillation were 1.9, 3.5, 4.4, and 6.5 Hz for freestream velocities of 1, 1.62, 2, and 3 m/s, respectively.

6.3.3 *Influence of the Turbine Moment of Inertia in the Runaway Angular Velocity*

Figure 6.6 shows the evolution of the angular velocity for four different moments of inertia of the turbine. First, it is clear that the overshoot in the evolution of the angular velocity decreases as the moment of inertia increases. It is also clear that as the moment of inertia increases the settling time increases, but the average runaway angular velocity is the same for the four different turbines. Settling times of 0.8, 1.0, 1.7, and 2.9 s were observed for moment of inertia of 0.5, 1.3, 3.0, and 8.9 kg m^2, respectively. As the moment of inertia increases, it is also observed that amplitude of the periodic behavior of the angular velocity decreases so that the operation of the turbine is smoother.

There are also some differences in the topology of the self-starting curve of the four turbines as the moment of inertia increases for example for $I = 8.9\,\text{kg m}^2$ and $I = 3.0\,\text{kg m}^2$ some wiggles are observed while for the other two the curve is more straight. Similar tendencies in the self-starting curve of a vertical axis wind turbine were observed by Untaroiu et al. [4].

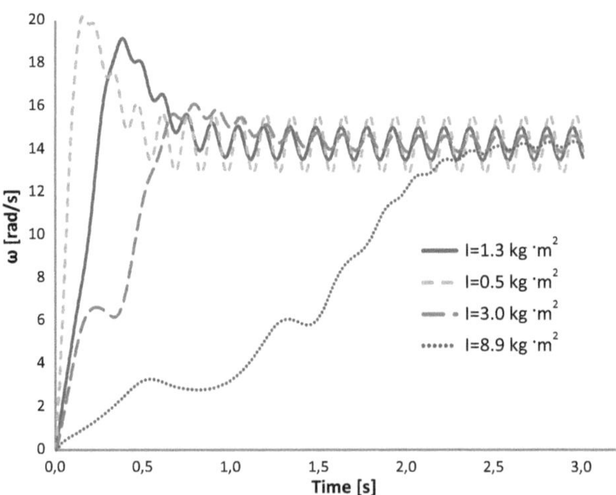

Fig. 6.6 Angular velocity evolution for different moment of inertia

6.4 Conclusions

A computational study of the transient start-up of a Darrieus type water turbine was presented and discussed. For this, a computational model was implemented in the commercial CFD software ANSYS-FLUENT v14, in which the solution of the Navier–Stokes equations was coupled with the RBD equation of the turbine. This coupling was achieved with the development and implementation of a UDF. Once the model was validated, a series of computational experiments were performed in order to study the influence of the moment of inertia and the freestream velocity in the dynamic performance of the turbine, in special the runaway angular velocity and the self-starting capabilities. Numerical results show that as the freestream velocity is increased, the runaway angular velocity of the turbine increases, which is consistent with the observation that the frequency of oscillation of the angular velocity (in the quasi steady-state) increases as the freestream velocity also increases. For a given turbine, it was observed that the increment in the moment of inertia of the turbine does not influence the average value of the runaway angular velocity (quasi steady state) but causes an increase in the time taken for achieving this quasi steady-state. Regarding the self-starting curve, some differences in the topology of this curve were observed of the four turbines as the moment of inertia increases, similar observations can be found in the literature for vertical axis wind turbines [4]. The proposed model has the capabilities to test several operation and transient conditions of the turbine, so it constitutes an interesting design tool.

Acknowledgements This project was sponsored by the Young Researchers program from the Colombian Administrative Department of Science, Technology and Innovation (Colciencias).

References

1. Lain S, Osorio C (2010) Simulation and evaluation of a straight-bladed Darrieus-type cross flow marine turbine. J Sci Ind Res 69:906–912
2. Ferreira C (2009) The near wake of the VAWT 2D and 3D views of the VAWT aerodynamics. PhD thesis, TU Delft
3. Lee Y (2013) Wind turbine simulation for time-dependet angular velocity, torque and power. Int J Eng Sci Technol 5:321–328
4. Untaroiu A, Wood H, Allaire P, Ribando R (2011) Investigation of self-starting capability of vertical axis wind turbines using computational fluid dynamics approach. J Sol Energy 133(4):041010 (8 pp.)
5. ANSYS ®Academic Research (2011) Release 14.0. ANSYS, Inc.
6. Belhache M, Guillou S, Grangeret P, Santa-Cruz A, Bellanger R (2013) Fluid structure interaction of a loaded Darrieus marine current turbine. Renew Energy Power Qual J 11. http://www.icrepq.com/icrepq%2713/515-belhache.pdf

7. Lain S, Lopez O, Quintero B, Meneses D (2013) Design optimization of a vertical axis water turbine with CFD. In: Ferreira G (ed) Alternative energies - advanced structured material, vol 34. Springer, Berlin/Heidelberg, pp 113–139
8. Dai Y, Lam W (2009) Numerical study of straight-bladed Darrieus-type tidal turbine. ICE-Energy 162:67–76
9. Gridgen Pointwise ® (2011) Release 15.17. Pointwise, Inc.

Chapter 7
The Physics of Starting Process for Vertical Axis Wind Turbines

Horia Dumitrescu, Vladimir Cardoş, and Ion Mălăel

Abstract In recent years there is a renewed interest in both large-scale and small-scale vertical axis wind turbines (VAWT). The development of multi-megawatt floating offshore VAWT is mainly a response to a plateau in the improvement of the aerodynamic performance of horizontal axis wind turbines. The research in small rotors (<100 kW) is motivated by the future demand for a decentralized sustainable energy supply in remote areas. However, such designs have received much less attention than the more common propeller-type designs and the understanding of some aspects of their operation remains, to this day, incomplete. This holds, because after some authors the starting operation is difficult, if not impossible, to induce the self-start capabilities without external assistance. This paper reviews the cause of the inability of the low solidity fixed pitch vertical axis wind turbines to self-start, and investigates the flow physics of dynamic stall in order to comprehend, interpret, understand and explain-all this comprising the problem of start-up.

7.1 Introduction

The aerodynamics of the lift-driven vertical axis wind turbine (VAWT) is referred to a rotational motion of the blades around an axis perpendicular to the flow direction, which is named Darrieus motion [1]. The energy extracted from the flow by the VAWT is the result of the work of the forces over the full rotation of the blade. The rotor configuration, composed of a number of blades with a chord-to-diameter ratio and blade aspect ratio, implies the azimuth variation of the bound circulation of the blade, with an inherently unsteady operation, resulting in a complicated wake, composed mainly of shed vorticity. Both the vorticity structure and the velocity/pressure field around the blade change depending on the operation regime.

H. Dumitrescu (✉) • V. Cardoş
"Gh. Mihoc – C. Iacob", Institute of Mathematical Statistics and Applied Mathematics, Calea 13 Septembrie 13, Bucharest 050711, Romania
e-mail: horiadumitrescu@yahoo.com

I. Mălăel
National Research & Development for Gas Turbines, COMOTI, Bucharest 061126, Romania

E. Ferrer, A. Montlaur (eds.), *CFD for Wind and Tidal Offshore Turbines*,
Springer Tracts in Mechanical Engineering, DOI 10.1007/978-3-319-16202-7_7

As the operation of VAWT implies the occurrence of all tip speed ratios from the start-up to the operating condition, a better understanding of the wake-flow physics and the blade–wake interaction in off-design operations is needed in order to achieve self-starting machines and minimizing their starting time.

The wake of a VAWT is typically divided into a near and a far wake. The near wake refers to the rotary flow field inside the rotor, which induces a velocity/pressure micro-flow field around the rotating blades. In the near wake, the presence of the rotor is apparent by the number of blades, chord-to-diameter ratio, blade aspect ratio, and blade aerodynamics, such as attached or stalled flows and 3D effects. The far wake is the region outside the rotor, where the actual rotor configuration is less important. The attentions for far wake flows are usually drawn in wake models, wake interference, turbulence models, and topographical effects. But a significant feature in the near wake of a VAWT is the vortical flow structures induced by the rotating blades during the starting process. The shed vorticity structure of the near wake has been found to change substantially as the rotor accelerates to its operational tip speed ratio. The evolution of the vorticity structures highly affects the behavior of wake-rotor system and the aerodynamic loads on the blades.

7.2 Start-Up of VAWT as a Self-Process

The lift-driven VAWT is, aerodynamically, a 2D wind-energy conversion system, where the energy conversion process occurs in the plane normal to the axis of rotation and in the volume generated by the revolution of a number of blades parallel to the rotation axis. The movement of the blades in a VAWT entails a variation of tangential and normal velocity perceived by the blade, resulting in a varying angle of attack and dynamics loads. At low tip speed ratios ($\lambda < 4$), the angle of attack of the blade exceeds the static stall angle, and this, in the case of unsteady flow, results in dynamic stall.

From the vectorial description of velocities, Fig. 7.1a, we can obtain the relative wind speed (W) and the expression that establishes the relationship between the angle of attack (AOA) α, the tip speed ratio (TSR) λ, and the azimuth angle θ of a blade in Darrieus motion (without velocity induction)

$$W = U_\infty \sqrt{\lambda^2 + 2\lambda \cos\theta + 1}, \tag{7.1}$$

$$\tan\alpha = \frac{U_\infty \sin\theta}{V_b + U_\infty \cos\theta} = \frac{\sin\theta}{\lambda + \cos\theta}\,. \tag{7.2}$$

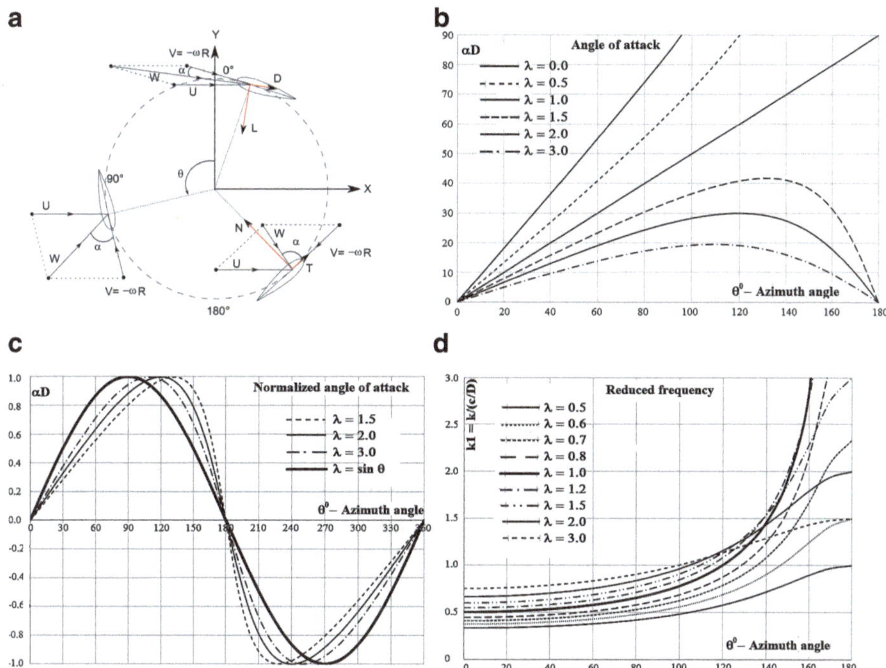

Fig. 7.1 Flow conditions for Darrieus motion: (**a**) typical flow velocities; (**b**) variation of angle of attack; (**c**) normalized angle of attack and sine-curve; (**d**) reduced frequency as a function of λ and the azimuth angle θ

Then, the reduced frequency (level of unsteadiness) can be expressed in terms of TSR as

$$k \equiv \frac{\omega c}{2W} = \left(\frac{c}{D}\right)\frac{\lambda}{\sqrt{\lambda^2 + 2\lambda\cos\theta + 1}}, \tag{7.3}$$

where c is the airfoil chord and D the diameter of the VAWT.

The variations of α, the normalized value $\alpha/\alpha_{\max}$, and the reduced frequency k are evaluated from Eqs. (7.2) and (7.3) as functions of the azimuth angle θ for different λ, and are shown in Fig. 7.1.

Figure 7.1 reveals that the Darrieus motion implies a wide range of operation regimes creating perturbations on the flow, both at the blade scale and the rotor scale, as:

– The variation of angle of attack for $\lambda \sim 1$ presents a strong discontinuity-like instability at the cyclic change of incidence direction ($\theta = 180°$), while for the values of $\lambda > 1$ the downwind passage of blades causes a weaker inflexional instability when the angle of attack becomes zero, Fig. 7.1b. Apart from these centrifugal instabilities the contribution of unsteadiness from the flow separation on blades is added at high angles of attack. The unsteady flow phenomenon produced by these cumulated perturbations of Darrieus motion is a complex physical mechanism involving the wake capture, concentration/relaxation of shed vorticity in the vicinity of the blade, detachment of concentrated vorticity-like vortex from the blade followed by the blade–vortex interaction, and the rotational circulation increase [2, 3].

- The variations of normalized angles of attack α at the higher values of λ ($\lambda \geq 2$) are very similar to sine-curve $\left(\alpha \approx \frac{\sin\theta}{\lambda}\right)$, with their peaks at about the same azimuth angle of 90°, while the variations at the low values of λ ($\lambda < 2.0$) contain elements of plunging motion with peaks at different azimuth angles, see Fig. 7.1c.
- The variation of reduced frequency k shows the existence of a band of tip speed ratios $\lambda \approx (0.7 - 1.5)$ with the rough increase of frequency like a discontinuity (Fig. 7.1d), which causes unsteady wakes at the rotor scale, resulting in structural modifications of wake and an efficient blade–wake interaction. At low tip speed ratios the Darrieus turbine operates with two distinct modes: mixed lift-drag driven and full-lift driven. The shift to full-lift driven state takes place when the tip speed ratio exceeds the value 1 and the alteration of the shed vorticity structure produces the continuous thrust production. The high unsteadiness levels promote to trigger off the structural modifications of wake and the efficient blade–wake interactions.
- For $\lambda \geq 4$ the variations of α are more significant at $\theta = 0°$ and $\theta = 180°$, where the shed vorticity of the wake is concentrated in these azimuth positions, consisting of two rows of small vortical structures. These vortices occurring in the downwind/upwind passage of the blade are the main sources of velocity induction which highly affect the downwind movement.

The unsteadiness/vorticity structure associated with the flow field and VAWT operating state can be classified into three levels (Fig. 7.1d):

- Zero level is the dispersed unsteadiness, when $k/(c/D) \leq 1.0$ and the tip speed ratio $\lambda \leq 0.5$ and $\lambda \geq 5.0$; commonly, its effect is neglected and a quasi-steady assumption is used for the flow around the airfoil/blade, while the unsteady wake contains shed vorticity with low degree of concentration (see Sect. 7.3).
- First level is the localized unsteady phenomenon of dynamic stall with lift increment at low angle of attack ($\alpha \sim 25°$), occurring at $\theta \sim 90°$ when $k/(c/D) \leq 2.0$ and $\lambda \geq 2.0$; its effect is heuristically simulated by the motion of a sinusoidal pitching airfoil [4] and the unsteady wake contains shed vorticity with middle degree of concentration.
- Second level is a localized unsteady phenomenon of dynamic stall with drag reduction at high angle of attack ($\alpha > 45°$) occurring in leeward region ($120° < \theta \leq 180°$) when $k/(c/D) > 2.0$ and $\lambda \approx 0.7 - 1.5$; its effect has been identified at low Reynolds numbers [3] where the unsteady wake contains shed vorticity with high degree of concentration.

Assuming that no power is extracted during starting and that the rotor torque Q acts only to accelerate the blades, the start-up is governed by the following equation [5]

$$\frac{d\lambda}{dt} = \frac{R}{JU_\infty} Q(\lambda), \tag{7.4}$$

where J is the total rotational inertia, Q is the driving torque and R is the rotor radius.

The differential equation (7.4) describing the Darrieus motion defines an autonomous dynamic system, nondependent explicitly on time, whose solutions are regular. This implies that except for the critical points where $\lambda = 0$, through every spinning point λ there exists one and only one path line, i.e. one regime. The dynamic system composed of rotor and wake creates cyclic perturbations by the rotating blades and unsteady flow within the wake, which in its turn induces a velocity/pressure field over blades resulting in the change of aerodynamic forces (C_L, C_D) and so on up to a operating condition is reached, according to the causality schema:

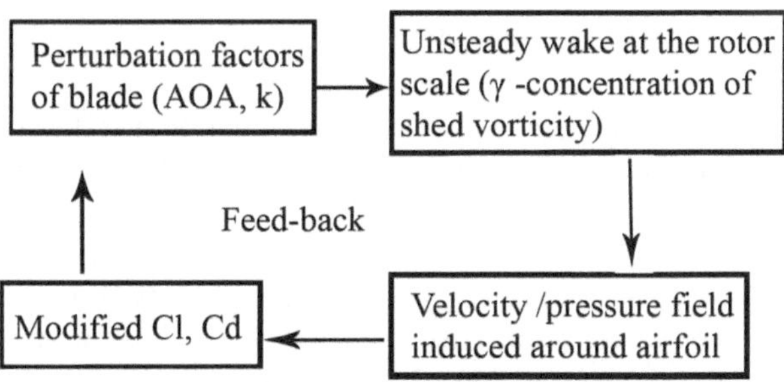

The physical mechanism of this system implying basically a linear-rotational flow transition phenomenon, consists in the successive alterations of the shed vorticity structure as the rotation increases, that can be triggered off only if a certain unsteadiness level of flow field is generated.

In contrast to the extremely intricate laminar-turbulent transition which is a concomitant rotational and dissipative phenomenon, as almost equal contributions, the linear-rotational flow transition is a simpler nonlinear interaction phenomenon dominated by strongly rotational effects, but which can be an interesting case for a transition phenomenon.

7.3 Stuart's Vorticity Model

The wake is a consequence of energy exchange on the rotor, consisting of a certain vorticity structure which induces a velocity/pressure field at the blade scale. As there are more operating modes it is evident that different vortical structures will also exist directly related to the physical process of the start/acceleration of the rotor.

On the other hand, the flow visualization, from computation or experimental PIV results, has found to be useful in widespread applications including in VAWTs. However, interpretation of flow visualization in unsteady flows has remained somewhat ambiguous. Central to all interpretations of flow visualization, but not yet clarified, is the degree of concentration of vorticity of the unsteady shear flow encountered in these applications. Here, the flow visualization of a mixing layer of $\tan h$ form, based on an exact solution of the Euler equations found by Stuart [6] is simulated numerically by tracking particles of the flow field.

Stuart's solution represents a spatially periodic mixing layer consisting of a single row of vortical structures. The change of the flow pattern is discussed by reference to the vorticity, which for two-dimensional motion is given by [6]

$$\zeta = -e^{-2\Psi} = -[C \cosh y + A \cos x]^{-2}, \tag{7.5}$$

where $\Psi = \ln(C \cosh y + A \cos x)$ is the stream function, x is the coordinate in the direction of the mean flow, and y is the coordinate normal to that direction; x and y velocity components are $u = \frac{\partial \Psi}{\partial y}$ and $v = -\frac{\partial \Psi}{\partial x}$.

Two important parameters are defined to describe the mixing layer: the extreme velocity ratio $R = (U_1 - U_2) / (U_1 + U_2)$, and the vorticity concentration parameter γ, which can take values from 0 to 1. If the velocity is nondimensionalized by the average velocity $\overline{U} = (U_1 + U_2)/2$, the streamwise and cross-stream velocity components can be expressed as

$$\overline{u} = 1 + R\frac{\sinh\ y}{\cosh\ y + \gamma \cos(x-t)}; \quad \overline{v} = R\frac{\gamma \sin(x-t)}{\cosh y + \gamma \cos(x-t)}. \tag{7.6}$$

As $y \to \infty$, $u \to 1 + R$, whereas when $y \to -\infty$, $u \to 1 - R$. When $R = 1$, only one stream is present. As the vorticity concentration parameter $\gamma \to 1$, a single row of point vortices is obtained, while for $\gamma \ll 1$, the unsteady shear flow corresponding to the linearized stability analysis of the $\tanh\ y$ profile is recovered. The dimensionless vorticity is given as:

$$\overline{\zeta} = -R\frac{(1-\gamma^2)}{(\cosh\ y + \gamma \cos(x-t))^2}. \tag{7.7}$$

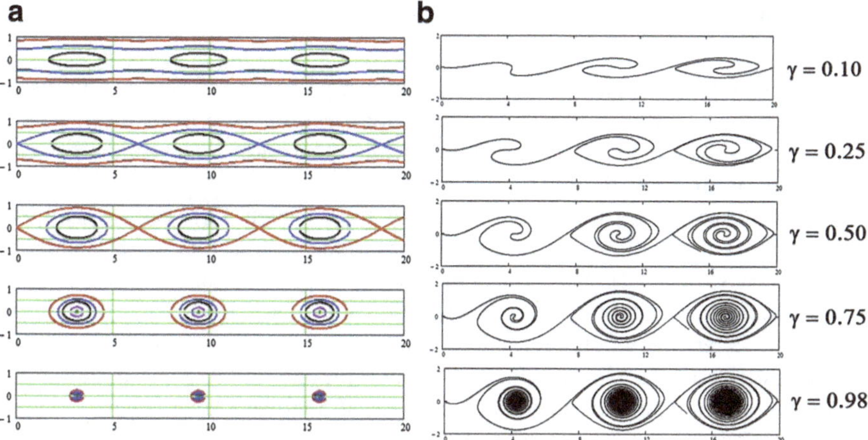

Fig. 7.2 (**a**) Contours of constant vorticity for several γ $\left(c\zeta/\overline{U} = -0.5, R = 1\right)$. (**b**) Effect of γ on streakline patterns

The vorticity distribution for $R = 1$ at time $t = 0$ is shown in Fig. 7.2a for several values of vorticity concentration parameter γ. For all vorticity contours, the outermost contour has the same value $\overline{\zeta} = -0.5$ (the increment between contour lines is $\delta\overline{\zeta} = -0.5$). The circulation of each vortical structure is $\Gamma = 4\pi R$ with $\Gamma = 4\pi$ for $R = 1$ (Fig. 7.2a).

From the vorticity distribution, time-averaged values at any point can be calculated. If we form the ratio of maximum total vorticity to maximum mean vorticity (at $y = 0$, $x - t = \pi$), we obtain

$$\frac{\overline{\zeta}_{\max}}{\overline{Z}_{\max}} = (1+\gamma)\sqrt{(1+\gamma)/(1-\gamma)}\,. \tag{7.8}$$

Equation (7.8) is a one-to-one relationship between this ratio and the vorticity concentration parameter γ, which can be used to estimate the vorticity concentration from the value $\frac{\overline{\zeta}_{\max}}{\overline{Z}_{\max}}$ found experimentally. There is a wide range of reported values of γ or $\frac{\overline{\zeta}_{\max}}{\overline{Z}_{\max}}$ in the literature, ranging from $\gamma = 0.25$ to $\gamma = 0.70$, depending upon the type of shear layer and the Reynolds number [7].

Furthermore, it is straightforward to visualize the flow using the tracking particles of the flow field via its material derivative to obtain the streakline patterns, Fig. 7.2b. As expected, the rate of roll-up of the particles increases with increasing concentration, and the cross-stream extent of the rolled-up particles approaches the same value for all cases, showing no dependency on circulation.

7.4 Rotational Effect on the Wake

The CFD simulation of airfoil flow with an AOA higher than 45° is rarely discussed in the literature. However, the blades encounter a very high AOA as they rotate at low TSR (as shown in Fig. 7.1b) and unsteady flow structures occur with large-scale separations around airfoils.

Prandtl described separation as a layer of fluid, which was set in rotation by the friction at the wall and continued its way into the free fluid as a Helmholtz discontinuity surface [8]. This concept evidently implies that the whole boundary layer no longer proceeds along the wall but transforms itself into a free vortex layer, and thereby alters the whole inviscid flow field. The above statement supposed the existence of a boundary layer, which is meaningful only in the approximation of a high Reynolds number, and hence this concept of boundary layer separation is an approximation. However, a more general concept of flow separation exists, i.e. that the fluid particles no longer proceed along the wall but turn into the interior of the flow field, which may be either outside or inside the rotational viscous fluid close to the body surface [9, 10]. The conception of fluid particle separation also includes large-scale separation at low Reynolds numbers—for instance, the separated flow around airfoils at high AOAs. In contrast to the boundary layer separation, fluid particle separation has nothing to do with the magnitude of the Reynolds number, and is a real physical situation found in experiments, which can be studied by either the Navier–Stokes equations (large-scale separation) or the boundary layer equation if only small-scale separation occurs inside the layer.

The particle flow separation conception and the Stuart's shed vorticity model can describe better the behavior of an unsteady wake, to which a rotor system is associated. Therefore, the applications of these two approaches are examined in the aerodynamic simulations of a single static airfoil and a straight-bladed VAWT [3].

The simulation of airfoil flow with high AOA has shown that the wake produces different particle path patterns as AOA increases, Fig. 7.3.

The increase of flow perturbation results in the accumulation of shed vorticity compressed behind the airfoil, which can be associated with the increment of its degree of concentration. Thus, the wake behind the airfoil behaves as a sinusoidal travelling wave or vorticity wave. The shape of the wake changes in terms of the angle of attack and the degree concentration for a constant strength of the shed vorticity. The change occurs as follows:

1. when $\alpha < \alpha_{ss}$, the boundary layer separates at the trailing edge, and free vorticity with low concentration ($\gamma < 0.25$) produces small vortices of order δ_l/c (boundary layer thickness) parallel to the streamline at trailing edge;
2. when $\alpha_{ss} \leq \alpha < 45°$, the flow separates at the leading edge, shed vorticity with middle concentration ($\gamma < 0.5$) accumulates and produces alternate vortex Karman street, *vortex size*/*c* < 1.0, see Fig. 7.3a;

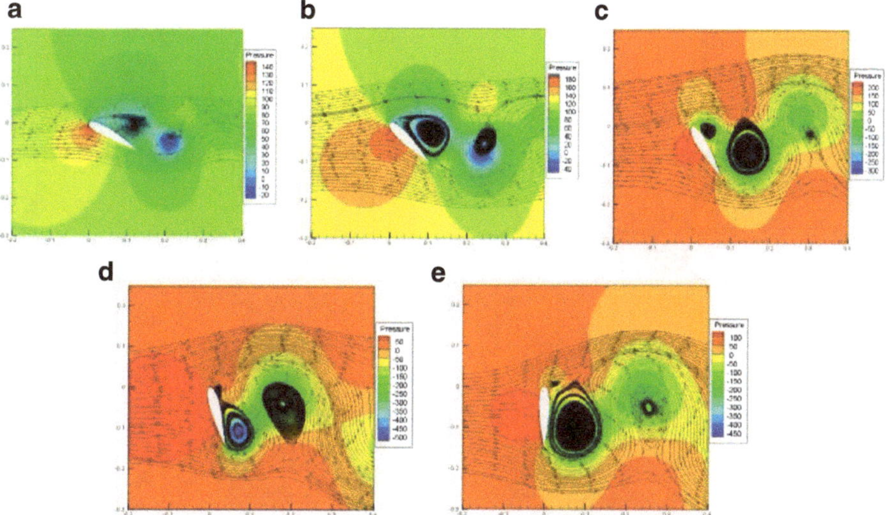

Fig. 7.3 Pressure field superimposed on the instantaneous streamlines for NACA0018 airfoil at $Re_c = 10^5$: single static airfoil. (**a**) $\alpha = 30°$; (**b**) $\alpha = 45°$; (**c**) $\alpha = 60°$; (**d**) $\alpha = 75°$; (**e**) $\alpha = 90°$

3. when $\alpha \geq 45°$, the vortex doublet/separation bubble separates from the airfoil, and conglomerates of highly concentrated shed vorticity ($\gamma \geq 0.5$) produce vortex doublet/separation bubble street, *vortex size/c* ≈ 1.0, see Fig. 7.3b–e. Considering Reynolds number circumstances used in [3], these concentrated vorticity structures are perceived as large separation bubbles shedding from the airfoil.

On the other hand, in contrast to the simple airfoil producing a free wake, the airfoil in Darrieus motion experiences a great influence due to the unsteady wake captured by rotor, as shown in Fig. 7.4 [3].

At low tip speed ratios, TSR ≈ 1.0, the near wake is captured inside the rotor and its shed vorticity is accumulated and concentrated in the windward quarter ($0° < \theta \leq 90°$) of the rotation, see Fig. 7.4a, b. Then in the leeward quarter ($90° < \theta \leq 180°$) of the rotation, the concentrated vorticity structure detaches from the airfoil and induces a velocity/pressure flow field over the own blade and previous blade resulting in the change of aerodynamic forces, see Fig. 7.4c–e. When the vorticity concentration/unsteadiness level of the flow field reaches a threshold-level ($\gamma \geq 0.5$), the tip speed ratio exceeds 1 and the continuous thrust production occurs, i.e. self-starting. For the lower unsteadiness levels the rotor is locked at low rotations and is not able to promote the continuous thrust-producing and self-starting.

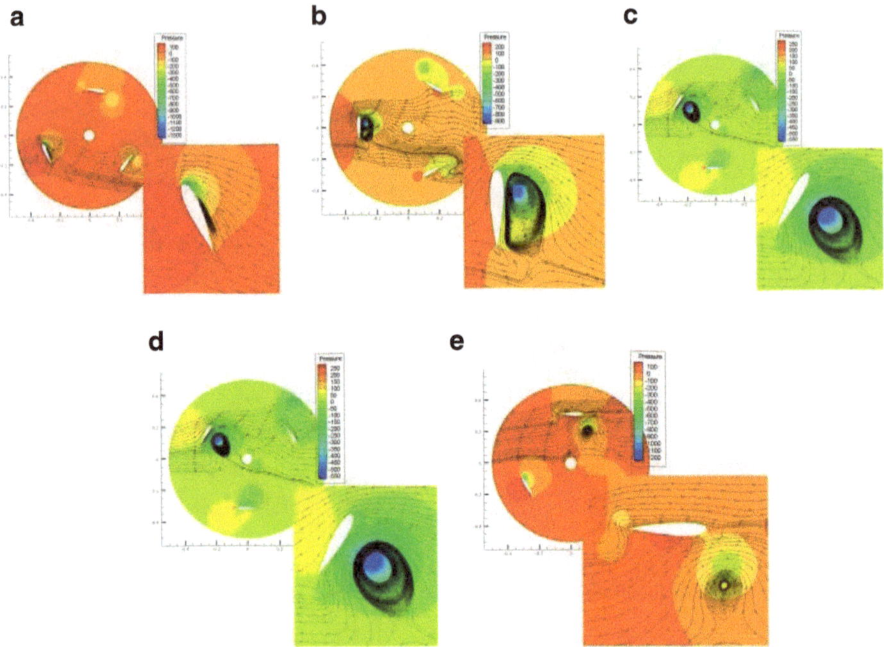

Fig. 7.4 Pressure field superimposed on the instantaneous streamlines for NACA0018 airfoil at $Re_c = 10^5$: 3-confined airfoils rotating at $\lambda = 1.0$. (**a**) $\alpha = 60°$; (**b**) $\alpha = 90°$; (**c**) $\alpha = 120°$; (**d**) $\alpha = 150°$; (**e**) $\alpha = 180°$

In the case when the rotor escapes from the dead band ($1 \leq \lambda \leq 1.5$), the concentrated shed vorticity relaxes/stretches and is recovered as bound circulation in the form of a leading-edge vortex. Concomitantly with the increase of the rotation, the concentration of shed vorticity decreases and disperses/breaks into smaller vortex structures, driven to rotor periphery by a centrifugal mechanism, according to the rule and Fig. 7.5 shown below.

$$\lambda\gamma = \text{const.}$$

The coupling of the rotational circulation ($\Gamma_R = \pi D \lambda U_\infty$) with the circulation Γ_S due to the shed vorticity can be achieved only for high values of the concentration degree of the vorticity ($\gamma \geq 0.5$), which are able to efficiently interact with the blade at the low TSRs.

Figure 7.6 illustrates this mechanism by means of the experimental PIV results at Reynolds $Re = 5 \cdot 10^4$ and $Re = 7 \cdot 10^4$, respectively for tip speed ratio $\lambda = 2$ and, $\lambda = 3$ and 4. Figure 7.6 shows the evolution of the clockwise vorticity shed from the airfoil after its rolling [11].

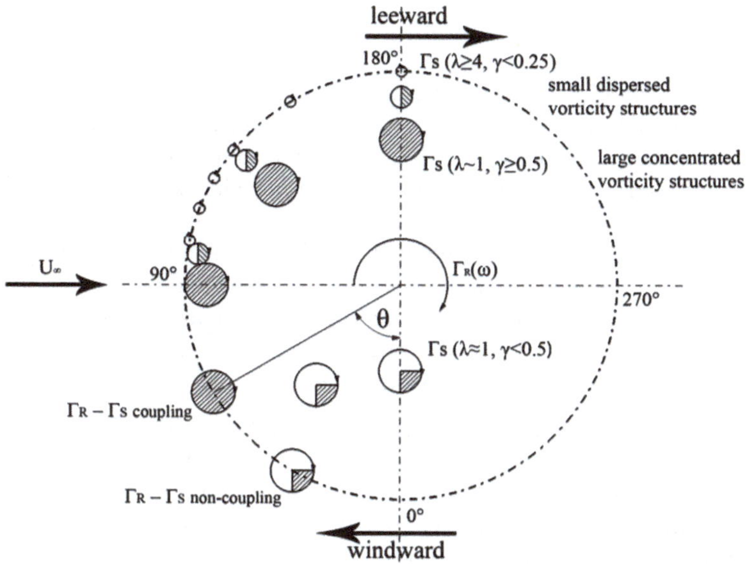

Fig. 7.5 Sketch of centrifugal mechanism for shed vorticity

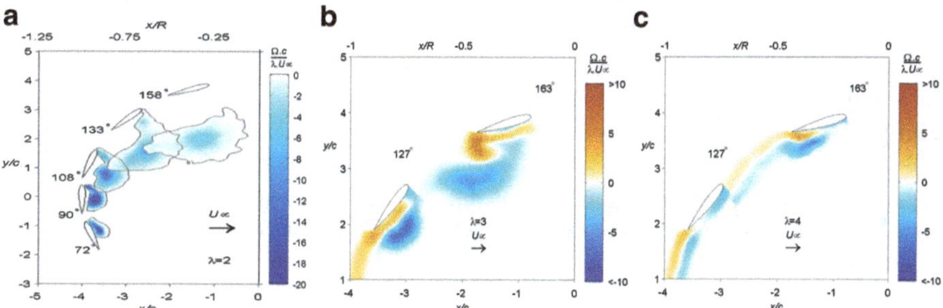

Fig. 7.6 Dispersion of the clockwise vorticity shed (*blue*) after the roll-up of leading edge: (**a**) $\lambda = 2,\ \gamma_{max} \approx 0.40$; (**b**) $\lambda = 3,\ \gamma_{max} \approx 0.30$; (**c**) $\lambda = 4,\ \gamma_{max} \approx 0.20$

The vorticity is dragged along with the airfoil, resulting in blade–vortex interaction phenomena which affect the pressure distribution and forces in the downwind passage of the blade. The differences from these cases lie in the concentration of the shed vorticity (but not in its magnitude) and the dispersion rate of the shed vorticity which increases at higher tip speed ratios. These results confirm our assumption that the occurrence and the magnitude of the dynamic stall event affect the distribution of shed vorticity at the rotor scale, thus resulting in conglomerates of shed vorticity concentrated in the upwind region of the flow field and a different induction field.

7.5 Conclusions

The starting of VAWTs is described as a transition process from linear flow, at low TSR, to rotational flow, at high TSR, which involves the change of vortical structure of wake and blade–wake interaction. As a perturbed autonomous system, once the system composed of the VAWT is initiated the machine will go faster up to the operating condition.

Based on Stuart's vorticity model and CFD computations, the mixing of shear layers in wake is devised as a two-step process: accumulation and concentration of shed vorticity when TSR is of order unity followed by relaxation of vorticity as the rotation increases and its recovery in bound circulation, in the form of a leading-edge vortex. Corresponding to the two steps there are different dynamic stall events:

1. a first dynamic stall event, named drag dynamic stall, is identified on confined-rotor blades at high AOAs generating significant drag reduction when TSR $\approx$ 1.0;
2. a second dynamic stall event, named lift dynamic stall, occurring at TSR $\approx$ 2–4 is apparently similar to a sinusoidal pitching airfoil and generates a substantially larger lift at post-stall AOAs and rotation increase by the relaxation of concentrated vorticity.

To achieve the self-starting of a VAWT at a low Reynolds number, the intrinsic unsteadiness generated by the instabilities of Darrieus motion-needs to be exploited. In other words, the onset of starting process is determined by a threshold-level of unsteadiness inside the rotor which can be achieved by proper sizing of the rotor configuration.

Acknowledgments This work was realized through the Partnership programme in priority domains—PN II, developed with support from ANCS CNDI—UEFISCDI, project no. PN-II-PT-PCCA-2011-32-1670.

References

1. Allet A, Halle S, Paraschivoiu I (1999) Numerical simulation of dynamic stall around an airfoil in Darrieus motion. J Solar Energy 121:69–76
2. Ferreira CJD, van Zurjlen A, Bily H, van Bussel G, van Kuik G (2010) Simulating dynamic stall in a two-dimensional vertical-axis wind turbine: verification and validation with particle image velocimetry data. Wind Energy 13:1–17
3. Malael I, Dumitrescu H, Cardos V (2014) Numerical simulation of vertical axis wind turbine at low speed ratios. Glob J Res Eng 1 Numer Methods 14(1):9–20
4. Wang S, Ingham DB, Lin M, Pourkashanian M, Zhi T (2010) Numerical investigations on dynamic stall of low Reynolds number flow around oscillating airfoils. Comput Fluids 39:1529–1541
5. Wood D (2011) Small Wind Turbines: analysis, design and application. Springer, London
6. Stuart JT (1967) On finite amplitude oscillations in laminar mixing layers. J Fluid Mech 29(3):417–440

7. Gursul I, Rockwell D (1991) Effect of concentration of vortices on streakline patterns. Exp Fluids 10:294–296
8. Goldstein S (1969) Fluid mechanics in the first half of the XXth century. Ann Rev Fluid Mech 1:1–28
9. Wu JZ, Gu JW, Wu JM (1988) Steady three-dimensional fluid particle-separation from arbitrary smooth surface and formation of free vortex layers. Z Flugwiss Weltraumforsch 12:89–98
10. Dallman U, Schewe G (1987) On topological changes of separating flow structures at transition Reynolds numbers. AIAA Pap 8:7–1266
11. Ferreira CS (2009) The near wake of the VAWT: 2D and 3D views of the VAWT aerodynamics. PhD Thesis, Delft University of Technology

Chapter 8
Hybrid Mesh Deformation Tool for Offshore Wind Turbines Aeroelasticity Prediction

Sergio González Horcas, François Debrabandere, Benoît Tartinville, Charles Hirsch, and Grégory Coussement

Abstract This paper describes a new development aiming to deform multi-block structured viscous meshes during fluid–solid interaction simulations. The focus is put on the deformation of external aerodynamic configurations accounting for large structural displacements and 3D multi-million cells meshes. In order to preserve the quality of the resulting mesh, it is understood as a fictitious continuum during the deformation process. Linear elasticity equations are solved with a multigrid and parallelized solver, assuming a heterogeneous distribution of fictitious material *Young modulus*. In order to improve the efficiency of the system resolution an approximate initial solution is obtained prior to the elastic deformation, based on *Radial Basis Functions* and *Transfinite interpolators*. To validate the performances of the whole algorithm, the DTU-10MW reference offshore wind turbine described by Bak et al. is analyzed (Description of the DTU 10 MW reference wind turbine. Technical report. Technical University of Denmark Wind Energy, Roskilde, 2013).

8.1 Introduction

Strong dynamical effects are expected during the whole life cycle of Offshore Wind Turbines (OWTs). The combination of this violent loading scenario and the slenderness of rotor blades requires considering Fluid–Structure Interaction (FSI) effects at the design stage.

Industry standards for OWT aerodynamic loads computations are based on simplified engineering models, Heege et al. [7], Jonkman and Buhl [12]. Even if these numerical approaches are less computationally demanding, three-dimensional flow effects are just estimated based on empirical corrections. Hence, the development of

S.G. Horcas (✉) • F. Debrabandere • B. Tartinville • C. Hirsch
NUMECA International, 189 Ch. de la Hulpe, 1170 Bruxelles, Belgium
e-mail: sergio.gonzalez@numeca.be

G. Coussement
Faculty of Engineering, Fluids-Machines Department, University of Mons, 53 Rue du Joncquois, 7000 Mons, Belgium

E. Ferrer, A. Montlaur (eds.), *CFD for Wind and Tidal Offshore Turbines*, Springer Tracts in Mechanical Engineering, DOI 10.1007/978-3-319-16202-7_8

more sophisticated CFD techniques for the detailed design of the rotor is justified as also stated by Li et al. [15].

To account for FSI phenomena, the discretized fluid domain needs to be re-adapted to structural deformations, motivating the implementation of *mesh deformation algorithms*. We have developed a mesh deformation tool based on the original formulation of the *Elastic analogy* described by Jasak and Weller [11]. The performances of the implemented methodology are compared with other approaches, such as the *Elliptic smoothing*. Special attention was paid to the deformation of complex meshes including tens of million points.

To illustrate the performances of the developed solution in the framework of a complete FSI simulation the DTU-10MW Reference Wind Turbine was studied, see Bak et al. [2]. All presented simulations were performed with the help of the commercial CFD package FINE™/Turbo, NUMECA International [17]. As a first approach, in order to focus on the study of rotor aeroelastic effects, the tower was not modeled in this preliminary analysis. A reduced-order model (ROM) was chosen for the structure, described by its mode shapes and natural frequencies. Steady FSI simulations at different operating points were performed, in order to quantify the expected deformations of the rotor.

8.2 Methodology

The commercial package FINE™/Turbo is used for this study. The flow solver is a three-dimensional, density-based, structured, multi-block Navier–Stokes code using a finite volume method. Central-difference space discretization is employed for the spatial discretization with Jameson type artificial dissipation. A four-stage explicit Runge–Kutta scheme is applied for the temporal discretization. Multigrid method, local time-stepping and implicit residual smoothing are used in order to speedup the convergence. Unsteady computations are performed using the dual time stepping approach described by Jameson [10]. For FSI applications, the CFD computational domain deforms according to the solid displacement. Additionally, it requires the resolution of Arbitrary Lagrangian–Eulerian (ALE) formulation of the Navier–Stokes equations, see Debrabandere [5].

The structure is represented by its natural frequencies and mode shapes. These are determined outside the flow solver and prior to any CFD computation, either by computation with a FEM structure solver or by experiments. Using these structural properties, the elastic body deformation under the action of the fluid loads is computed by a structural solver integrated inside the flow solver, Debrabandere [5].

As the mode shapes are defined on a *Finite Element* mesh, some interpolation issues between structure and fluid data may occur, see Fenwick and Allen [6]. In order to avoid this, the mode shapes are interpolated onto the fluid mesh prior to the coupled computation as suggested by Sayma et al. [18].

8.3 Mesh Deformations Methods Overview

Several mesh deformation tools are already available in the described code. In particular, a point-by-point scheme based on *Radial basis function* (RBF) interpolation developed by De Boer et al. [4] has been proved as a robust and high quality tool for a wide variety of applications.

RBF method uses the displacement of the boundary nodes in order to construct an interpolation function $u(\mathbf{x}_{\text{ref}})$ as a sum of *RBFs*:

$$\mathbf{x} = \mathbf{x}_{\text{ref}} + u(\mathbf{x}_{\text{ref}}) = \mathbf{x}_{\text{ref}} + \left[\sum_{b=1}^{n_b} \alpha_b \phi \left(\| \mathbf{x}_{\text{ref}} - \mathbf{x}_{\text{ref},b} \| \right) + p\left(\mathbf{x}_{\text{ref}}\right) \right], \tag{8.1}$$

With:

- $\mathbf{x}$ the deformed mesh node position;
- $\mathbf{x}_{\text{ref}}$ the original mesh node position;
- $\mathbf{x}_{\text{ref},b}$ the original mesh position of boundary node b;
- n_b the total number of boundary nodes;
- α_b interpolation coefficients, determined according to boundary nodes displacement;
- ϕ the *RBF* (in this case, *Thin Plate spline*);
- $p(\mathbf{x}_{\text{ref}})$ a first order polynomial;

Figure 8.1 illustrates the performances of this approach when dealing with imposed large displacements in a 2D vortex induced vibrations test case [9] for an 8×10^3 nodes mesh. This method is however limited by its scalability. As stated by Bos et al. [3], the cost of calculation of the interpolation coefficients scales with $\mathcal{O}(\text{n}_\text{b}{}^3)$ and the new coordinates evaluation with $\mathcal{O}(\text{n}_\text{b}\text{n}_\text{i})$, where n_i is the number of inner mesh nodes. This is the reason why it is hardly applicable to multi-million 3D meshes.

In order to develop a fast and robust method for big OWT rotor blades mesh deformation, a connectivity-based approach was chosen from the very beginning. In this line, a popular and simple technique is the so-called *Elliptic smoothing*,

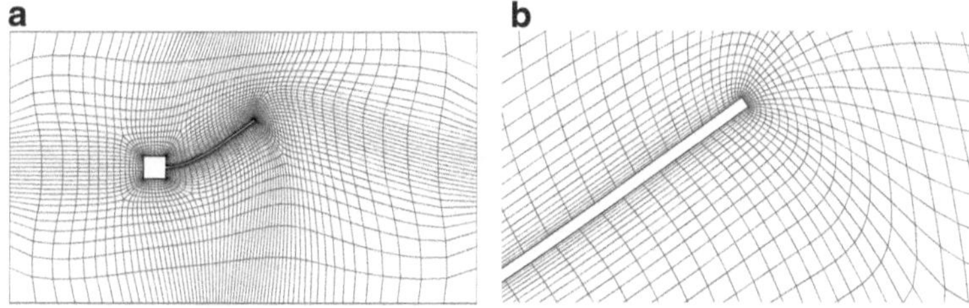

Fig. 8.1 Vortex induced vibrations 2D test case, *radial basis functions* deformation. (**a**) Global view. (**b**) Beam tip detail

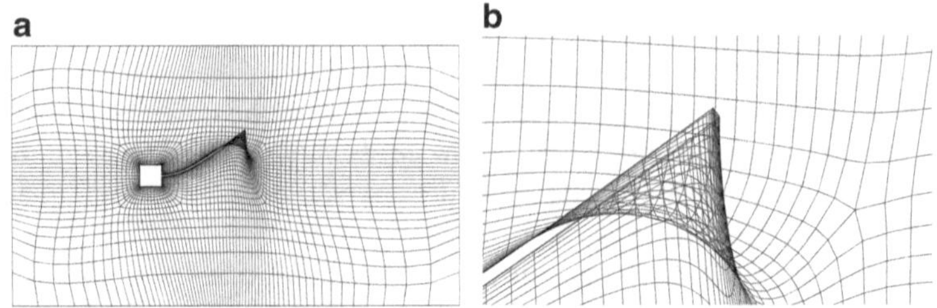

Fig. 8.2 Vortex induced vibrations 2D test case, *Elliptic smoothing* deformation. (**a**) Global view. (**b**) Beam tip detail

which was implemented in the considered code prior to this research, see NUMECA International [17]. With this method, the new mesh nodes position is calculated by solving the linear system:

$$\nabla \cdot (\omega \nabla (\mathbf{x} - \mathbf{x}_{\text{ref}})) = 0 \, , \tag{8.2}$$

where ω is a local diffusivity factor aiming to preserve the final mesh quality by limiting the deformation of small volume cells.

Even if the diffusivity is controlled by ω, the limitations of the *Elliptic smoothing* when dealing with large displacements have been extensively described in the literature: Hermansson and Hansbo [8], Karman [13], Arabi et al. [1]. In particular, mesh folding is often observed around concave regions as illustrated in Fig. 8.2 (same views as Fig. 8.1 are kept). The new developed algorithm is based on the *Elastic Analogy* described by Jasak and Weller [11]. In this case, the diffusion mechanism is controlled by considering the CFD mesh as an elastic continuum. A finite volume discretization has been coded in order to reproduce the elastic equations of our fictitious material:

$$\int\int_S \left(\lambda \, (\nabla \cdot \mathbf{x}) \, \mathbf{I} + \mu \left((\nabla \mathbf{x})^T + \nabla \mathbf{x}\right)\right) \cdot \mathbf{n} \, d\mathbf{S} = 0 \, , \tag{8.3}$$

where $\mathbf{I}$ refers to the unit tensor, and λ and μ correspond to the so-called *Lame Coefficients* and can be directly related to the elastic properties (*Young Modulus* and *Poisson Ratio*) given to our continuum:

$$\mu = \frac{E}{2\,(1+\nu)}, \tag{8.4}$$

$$\lambda = \frac{\nu E}{(1+\nu)\,(1-2\nu)}, \tag{8.5}$$

In order to control the orthogonality of our mesh near the walls and its overall quality, a heterogeneous distribution of the elastic properties is needed, see Stein

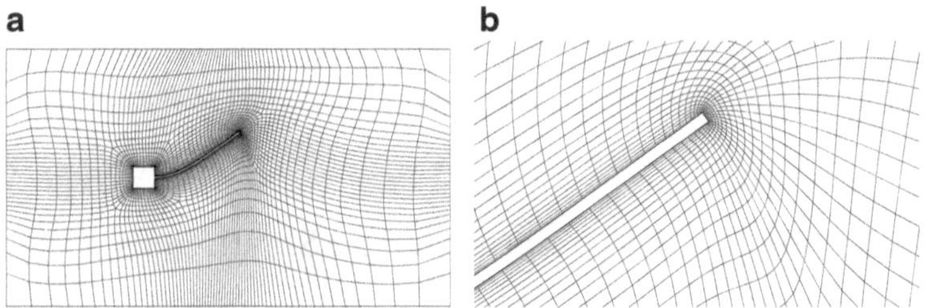

Fig. 8.3 Vortex induced vibrations 2D test case, *Elastic analogy* deformation. (**a**) Global view. (**b**) Beam tip detail

et al. [21], Yang and Mavriplis [22]. A good resulting mesh quality can then be obtained for the considered 2D example, as seen in Fig. 8.3. In the framework of structured meshes, a very interesting tool to be used in mesh deformation is the *TransFinite Interpolation* (TFI). This method cannot be considered as a mesh deformation technique by its own since it needs as an input the displacements of our blocking topology. However, its reduced computational time makes it very attractive as an intermediate or final step in the complete mesh deformation process.

The arc-length based TFI method suggested by Lai et al. [14] has been implemented for this purpose. Let us consider a mesh segment going from point A to B. Then we can define the arc-length of an interior point i as:

$$l_i = \sum_{j=A+1}^{i} |x_{\text{ref},j} - x_{\text{ref},j-1}| \,, \tag{8.6}$$

where x_{ref} corresponds to the first dimension of our nodes reference position vector $\mathbf{x}_{\text{ref}}$. We can also define a total arc-length of the segment as:

$$L_{BA} = \sum_{j=A+1}^{B} |x_{\text{ref},j} - x_{\text{ref},j-1}| \,, \tag{8.7}$$

Then, new edge interior points coordinates can be computed by performing a linear interpolation:

$$x_i - x_{\text{ref},i} = \left[1 - \frac{l_i}{L_{BA}}\right](x_A - x_{\text{ref},A}) + \left[\frac{l_i}{L_{BA}}\right](x_B - x_{\text{ref},B}) \,, \tag{8.8}$$

The same methodology is repeated in all the directions.

As performed by Lai et al. [14], this *1D-edges* interpolation can be easily extrapolated to a *2D-faces* version for block surface interpolation or to *3D-volumes* for block interior nodes recomputation. The combination of these 1D/2D/3D interpolators allows then to remap our blocks starting either from the deformed position of block corners, edges or faces. The main drawback of this approach is its inability to adequately maintain mesh quality through large mesh deformations. This is particularly true when structural deflections account for large rotations.

8.4 Validation Test Case, Aerodynamic Simulations

For the validation of the presented aeroelastic prediction methodology, the reference wind turbine described by Bak et al. [2] at different working points was analyzed, assuming a steady flow behavior.

Aerodynamic simulations were performed in order to validate our CFD case setup with respect to the available literature references. CFD RANS simulations at different 0° pitch operating points were carried out for this purpose.

8.4.1 Mesh Generation

In order to build a 3D mesh of the DTU-10MW wind turbine, blade sections were imported in AUTOGRID5™ from data published by Bak et al. [2]. Blade surfaces at 0° pitch, as well as nacelle and hub geometries were included in this process.

A blocking topology was established around the blade, putting special attention in the local mesh around the blunt edge and blade tip. Final mesh accounted for 7.2×10^6 nodes. A first cell size of 3.0×10^{-5} m was imposed around the considered geometry, in order to properly describe the boundary layer for the studied wind speed range.

8.4.2 CFD Results

Our first CFD simulation aimed to analyze the nominal operating point of the DTU-10MW reference wind turbine in a rigid rotor configuration:

- Rotational speed: 8.836 RPM
- Incoming wind speed: $11\ \mathrm{m\,s^{-1}}$
- Pitch angle: 0°

A Spalart–Allmaras turbulence model [20] was used for our RANS simulation, enhanced by Merkle preconditioner [16]. A multigrid approach was followed in order to speedup our computations, which were launched in parallel.

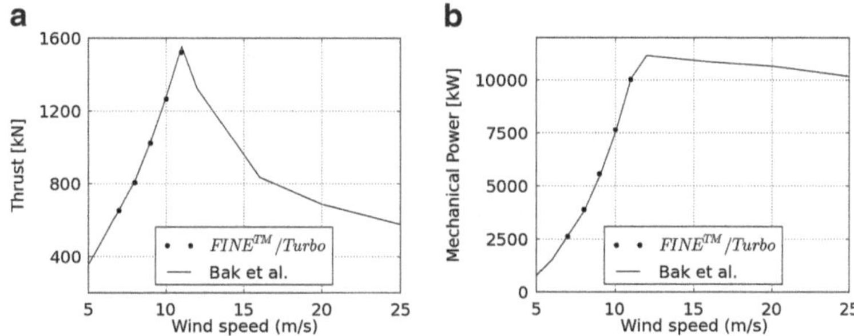

Fig. 8.4 DTU-10MW total rotor thrust and mechanical power versus wind speed. (**a**) Total thrust. (**b**) Mechanical power

Obtained total thrust and mechanical power were in agreement with the numerical results already published by Bak et al. [2], as well as span evolution of local thrust and power coefficients.

In order to extend the validation of our flow setup, different operating points of the DTU-10MW wind turbine at 0° pitch were also analyzed (Fig. 8.4).

8.5 Validation Test Case, Aeroelasticity Prediction

A steady aeroelastic simulation was performed starting from the already described CFD computation at the nominal operating point of the DTU-10MW reference wind turbine. The behavior of the structure was linearized by means of its natural frequencies and deformed shapes. An iterative procedure *fluid simulation/structure deformation* allowed us to find the final static deformed blade and flow behavior after several steps.

8.5.1 Structural Model

For the considered aeroelastic simulations, the structure was linearized by means of the ROM developed by Debrabandere [5].

A modal analysis was performed within the commercial package Abaqus [19], thanks to the model provided by Bak et al. [2]. Centrifugal effects were included in this computation in order to take into account their impact on blade stiffness and blade deformation at the considered rotational speed.

8.5.2 Mesh Deformation

A finite volume based version of the *Elastic Analogy* was used as the main mesh deformation technique for this challenging case. An under relaxed Jacobi solver was implemented and parallelized in order to efficiently solve our linear system. A multigrid strategy was setup, allowing us to reduce the computational time by accessing coarser grid levels of our mesh.

To increase the efficiency of our system resolution, a good initial solution was provided by means of a mix of other mesh deformation algorithms. The implemented procedure can be summarized as:

RBF — Mesh block corners placed at domain boundaries serve as a base for the *RBFs Interpolation* of interior mesh block corners. This topologically based nodes selection allows us to reduce the important computational time attached to RBF approach

TFI — Block edges, faces and interior points are computed based on *TFI*, in order to have a complete approximate mesh before solving our linear system

ELA — Interior block nodes are then updated by means of an *Elastic Analogy* based on the reference mesh in order to improve resulting mesh quality. A *Young modulus* proportional to the inverse of wall distance is used. The *Poisson Ratio* of the fictitious material is set to 0.25

8.5.3 Fluid–Solid Interaction Results

Nine hundred iterations of fluid simulation were performed. The structure deformation is first calculated at the beginning of the computation. Then it is updated twice during the computation at iterations 300 and 600 according to the fluid load evolution. Figure 8.5a shows that the computed total thrust is stabilized even after the second deformation (which does not considerably modify our blade shape). Obtained results matched with the trend already established in previous computations of Bak et al. [2], since the computed thrust is considerably lower when aeroelastic effects are taken into account. As expected, the resulting blade deformation is mainly dominated by first flapping mode. Figure 8.5b superposes initial and deformed blade geometries. Table 8.1 illustrates the performances of the hybrid mesh deformation method described in Sect. 8.5.2. Grid computed by the combination of RBF and TFI algorithms is already able to keep original mesh quality parameters, at a very low CPU cost. The additional computational effort of the ELA step of Sect. 8.5.2 is justified by looking at the local mesh quality around the blade. Near-wall orthogonality is highly improved by the introduction of the *Elastic Analogy* in spanwise direction (Fig. 8.6a–c). For the blade sections mesh, the absence of an important blade twist deflection allows the RBF+TFI approach

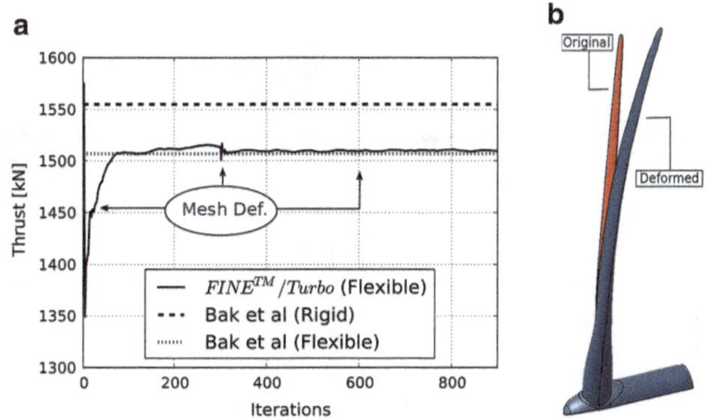

Fig. 8.5 Nominal operating point aeroelastic results of DTU-10MW. (**a**) Thrust evolution during FSI simulation. (**b**) Deformation

Table 8.1 Mesh quality of original and deformed DTU 10MW blade mesh

	OR (°)	AR	ER	Computational data		
Def. method	Min.	Max.	Max.	#Proc	r_{def}	Mem. [Mb]
Original	15.2	41,700	1.66	–	–	–
RBF+TFI	15.2	47,200	1.66	9	0.003	1,385
RBF+TFI+ELA	15.2	47,200	1.66	9	0.39	3,706

r_{def}: Ratio between mesh deformation and fluid simulation times
Mem.: Maximum memory allocated, divided by the number of processors (#Proc)
OR orthogonality, *AR* aspect ratio, *ER* expansion ratio

to provide high quality deformed grids (Fig. 8.6d–f). The blade tip displacement measured out of the rotor plane was 7.82 m. Tilt and cone angles (5° and 2.25°, respectively) were not taken into account in our simulation. We can however estimate the tip displacement in our final rotor configuration of 7.74 m, by projecting the computed deformation.

This displacement implies a margin of 58 % with respect to the tower clearance designed in Bak et al. [2]. As expected, this deformation is lower than the maximum tip deflection measured by the same authors in the *Power Production operation DLCs*, where it ranges from 10.5 to 12.4 m. In that analysis a wider variety of incoming wind speeds and directions were taken into account, including near cut-off wind speed configurations.

8.6 Conclusions and Future Works

In order to predict aeroelastic effects of multi-megawatt OWTs rotor, a numerical approach has been developed. In particular, a new mesh deformation code based on the *Elastic Analogy* has been implemented. The performances of the method

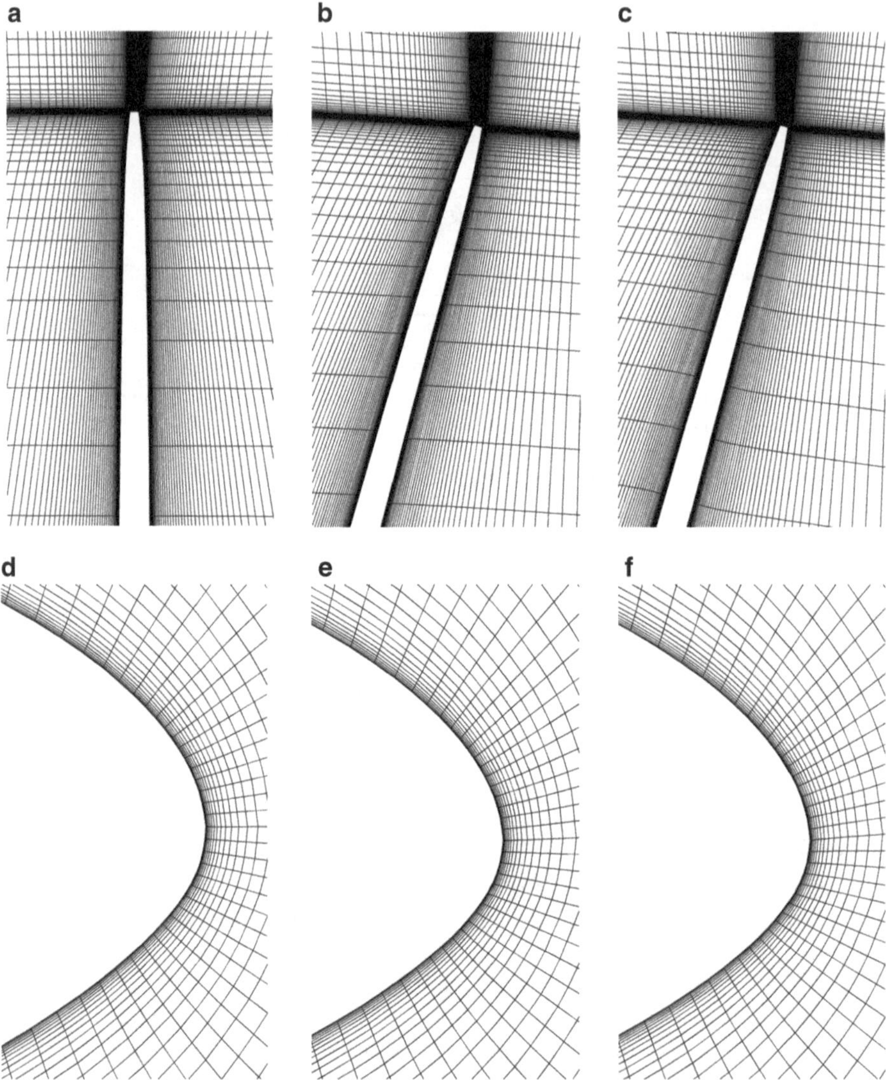

Fig. 8.6 DTU-10MW, original and deformed meshes. Details of spanwise cut (**a**)–(**c**) and leading edge at 90 % span (**d**)–(**f**)

when dealing with large geometrical deflections have been improved, thanks to the coupling with two other complementary approaches: the *TFI* and the *RBF*.

The efficiency and robustness of the algorithm have been evaluated at the nominal operating point of the DTU-10MW reference wind turbine. Both aerodynamics and aeroelastic results were successfully checked against previous numerical analyses performed by third party softwares. No experimental data are currently available for validation of the presented results.

Future works will be first devoted to extend the presented rotor flow setup to unsteady configurations. The actual mesh will be re-built in order to include a DTU-10MW tower model. The interaction between flexible rotor and tower will be then characterized. After the consolidation of this unsteady aeroelasticity prediction methodology, the model will serve as a basis for a set of new developments in the code, aiming to account for hydrodynamic excitations on bottom-fixed and floating OWT configurations.

Acknowledgements The authors acknowledge the European Commission (EC) for their research grant under the project FP7-PEOPLE-2012-ITN 309395 MARE-WINT (*new MAterials and REliability in offshore WINd Turbines technology*), see: http://marewint.eu/.

References

1. Arabi S, Camarero R, Guibault F (2012) Unstructured mesh motion using sliding cells and mapping domains. In: 20th annual conference of the CFD Society of Canada
2. Bak C, Zahle F, Bitsche R, Kim T, Yde A, Henriksen LC, Natajaran A, Hansen MH (2013) Description of the DTU 10 MW reference wind turbine. Technical report. Technical University of Denmark Wind Energy, Roskilde. http://dtu-10mw-rwt.vindenergi.dtu.dk
3. Bos FM, van Oudheusden BW, Bijl H (2013) Radial basis function based mesh deformation applied to simulation of flow around flapping wings. Comput Fluids 79:167–177. ISSN 00457930. doi:10.1016/j.compfluid.2013.02.004
4. De Boer A, Van der Schoot M, Bijl H (2007) Mesh deformation based on radial basis function interpolation. Comput Struct 85(11–14):784–795. ISSN 00457949. doi:10.1016/j.compstruc.2007.01.013
5. Debrabandere F (2014) Computational methods for industrial fluid-structure interactions. PhD thesis, Université de Mons (UMONS)
6. Fenwick CL, Allen CB (2007) Flutter analysis of the BACT wing with consideration of control surface representation. In: 25th AIAA applied aerodynamics conference
7. Heege A, Gaull A, Horcas SG, Bonnet P, Defourny M (2013) Experiences in controller adaptations of floating wind turbines through advanced numerical simulation. In: AWEA wind power conference, Chicago
8. Hermansson J, Hansbo P (2003) A variable diffusion method for mesh smoothing. Commun Numer Methods Eng 19(11):897–908. ISSN 10698299. doi:10.1002/cnm.639
9. Hübner B, Walhorn E, Dinkler D (2004) A monolithic approach to fluid-structure interaction using space-time finite elements. Comput Methods Appl Mech Eng 193(23–26):2087–2104. ISSN 00457825. doi:10.1016/j.cma.2004.01.024
10. Jameson A (1991) Time dependent calculations using multigrid, with applications to unsteady flows past airfoils and wings. In: 10th computational fluid dynamics conference. Fluid dynamics and co-located conferences. American Institute of Aeronautics and Astronautics, Honolulu. doi:10.2514/6.1991-1596
11. Jasak H, Weller HG (2000) Application of the finite volume method and unstructured meshes to linear elasticity. Int J Numer Methods Eng 48:267–287
12. Jonkman JM, Buhl ML Jr (2007) Development and verification of a fully coupled simulator for offshore wind turbines. In: NREL/CP-500-40979. National Renewable Energy Laboratory
13. Karman SL Jr (2010) Virtual control volumes for two-dimensional unstructured elliptic smoothing. In: 19th international meshing roundtable (IMR), pp 121–142

14. Lai KL, Tsai HM, Liu F (2003) Application of spline matrix for mesh deformation with dynamic multi-block grids. In: 21st AIAA applied aerodynamics conference. doi:10.2514/6.2003-3514
15. Li Y, Paik KJ, Xing T, Carrica PM (2012) Dynamic overset CFD simulations of wind turbine aerodynamics. Renew Energy 37(1):285–298. ISSN 09601481. doi:10.1016/j.renene.2011.06.029
16. Merkle CL, Sullivan JY, Buelow PEO, Venkateswaran S (1998) Computation of flows with arbitrary equations of state. AIAA J 36(4):515–521. ISSN 0001-1452. doi:10.2514/2.424
17. NUMECA International (2013) FINE™/Turbo v9.0 user manual
18. Sayma A, Vahdati M, Imregun M (2000) An integrated nonlinear approach for turbomachinery forced response prediction. Part I: Formulation. J Fluids Struct 14(1):87–101. ISSN 08899746. doi:10.1006/jfls.1999.0253
19. Simulia DSC (2008) Abaqus Analysis version 6.8 user's manual
20. Spalart P, Allmaras S (1992) A one-equation turbulence model for aerodynamic flows. In: 30th aerospace sciences meeting and exhibit. Aerospace sciences meetings. American Institute of Aeronautics and Astronautics, Reno. doi:10.2514/6.1992-439
21. Stein K, Tezduyar T, Benney R (2003) Mesh moving techniques for fluid-structure interactions with large displacements. ASME J Appl Mech 70(1):58. ISSN 00218936. doi:10.1115/1.1530635
22. Yang Z, Mavriplis DJ (2005) Unstructured dynamic meshes with higher-order time integration schemes for the unsteady Navier-Stokes equations. In: 41th aerospace sciences meeting and exhibit. American Institute of Aeronautics and Astronautics, Reno

Chapter 9
Numerical Simulation of Wave Loading on Static Offshore Structures

Hrvoje Jasak, Vuko Vukčević, and Inno Gatin

Abstract This chapter presents numerical simulations of water waves using the Finite Volume Method. Wave loads exerted on a truncated circular cylinder are calculated and compared to experimental data.

Mathematical model of two-phase incompressible flow is based on Navier–Stokes equations with Volume-Of-Fluid method for interface capturing. Waves are generated and absorbed using relaxation zones with prescribed potential flow solution. Potential flow and CFD solutions are blended implicitly within governing equations. The novel approach allows stable simulations at higher Courant–Friedrichs–Lewy numbers. The algorithm is described in detail and implemented within OpenFOAM/FOAM-extend in `Naval Hydro` pack.

The method is validated on two test cases, both regarding truncated circular cylinder. The first test case considers maximum regular wave loads with different frequencies and wave heights. The second test case simulates phase focused freak wave and its impact on the cylinder.

9.1 Introduction

The design of offshore structures requires information about peak loads that might occur during their service. The safety margin depends on the reliability of methods used to calculate these loads: smaller safety factors result in lower production and operational cost of structures. In order to acquire more precise information about wave loads, Computational Fluid Dynamic (CFD) methods are getting more

H. Jasak (✉)
Wikki Ltd, 459 Southbank House, SE1 7SJ London, UK

Faculty of Mechanical Engineering and Naval Architecture, University of Zagreb, Ivana Lučića 5, 10000 Zagreb, Croatia
e-mail: h.jasak@wikki.co.uk; hrvoje.jasak@fsb.hr

V. Vukčević • I. Gatin
Faculty of Mechanical Engineering and Naval Architecture, University of Zagreb, Ivana Lučića 5, 10000 Zagreb, Croatia
e-mail: vuko.vukcevic@fsb.hr; innogatin@gmail.com

E. Ferrer, A. Montlaur (eds.), *CFD for Wind and Tidal Offshore Turbines*,
Springer Tracts in Mechanical Engineering, DOI 10.1007/978-3-319-16202-7_9

attention. In [4, 5] active wave generation and absorption is described with moving boundaries and tested on various cases relevant to coastal engineering. On the other hand, Monroy et al. [9] used the Spectral Wave Explicit Navier–Stokes Equations (SWENSE) method to calculate the forces on a buoy in both regular and irregular seas.

In this work, the Finite Volume Method (FVM) is used to calculate wave loads on static cylindrical structures. Such structures are often used as mounting points for offshore wind turbines (such as Tension Leg Platforms (TLP) and Spar Platforms). The phenomenon of freak wave which can be extremely large and steep, naturally occurring wave is especially dangerous for offshore structures and the calculation of freak wave loads is a challenging task using conventional methods. One such calculation is presented here.

This chapter is organized as follows. In Sect. 9.2 governing equations for two-phase flows are briefly described. Wave modelling using relaxation zones and its implicit treatment is the focal point of this section. In Sect. 9.3 some details of the numerical procedure are given. In Sect. 9.4, considered test cases are presented. Finally, a short conclusion is given, discussing the results of simulations and future work.

9.2 Mathematical Model

The mathematical model of incompressible, two-phase flow is presented in this section. Introducing the Volume-Of-Fluid (VOF) method for interface capturing [12] allows formulation of single continuity and momentum equation for mixture of phases. Furthermore, wave modelling using relaxation zones [6] is described. Finally, implicit blending technique is presented, where the CFD solution is forced to correspond to the potential flow solution within governing equations in order to efficiently generate and absorb waves.

9.2.1 Continuity and Momentum Equations

Numerical simulations are based on continuity (9.1) and mixture momentum equation (9.2). For viscid, incompressible and Newtonian fluid within mixture model, the volumetric continuity equation can be written as:

$$\nabla \cdot \mathbf{u} = 0\,, \tag{9.1}$$

and the mixture momentum equation reads:

$$\frac{\partial(\rho\mathbf{u})}{\partial t} + \nabla \cdot (\rho\mathbf{u}\mathbf{u}) = -\nabla p_d + \nabla \cdot (\mu\nabla\mathbf{u}) + \nabla\mathbf{u} \cdot \nabla\mu - \mathbf{f} \cdot \mathbf{x}\nabla\rho + \sigma\kappa\nabla\alpha\,. \tag{9.2}$$

Here, $\mathbf{u}$ is the velocity field, p_d is the dynamic pressure from $p = p_d + \rho\mathbf{g} \cdot \mathbf{x}$ decomposition. α is the volume fraction. Gravitational force is denoted with $\mathbf{f}$. According to [2], the last term models the surface tension effects. Furthermore, ρ and μ present density and dynamic viscosity, respectively. For more details on the derivation of the above equation, the reader is referred to [11].

9.2.2 Volume of Fluid Equation

The method is based on the α indicator scalar field which is defined as:

$$\alpha(\mathbf{x}) = \begin{cases} 1\,, & \text{if } \mathbf{x} \in \Omega_1\,, \\ 0 < \alpha < 1\,, & \text{if } \mathbf{x} \text{ in the transitional area}\,, \\ 0\,, & \text{if } \mathbf{x} \in \Omega_2\,. \end{cases} \tag{9.3}$$

Ω_1 is the part of the domain occupied by the first phase (water), while Ω_2 is the part of the domain that contains the second phase (air). Cells that contain the free surface have value of α between 0 and 1. Hence, α represents the ratio of the volume of the first phase in given cell (V_1) and the total volume of the cell (V):

$$\alpha = \frac{V_1}{V}\,. \tag{9.4}$$

Fluid properties need to be defined before the solution of momentum equation, and are obtained using α:

$$\begin{aligned} \rho &= \alpha\rho_1 + (1-\alpha)\rho_2\,, \\ \mu &= \alpha\mu_1 + (1-\alpha)\mu_2\,. \end{aligned} \tag{9.5}$$

The transport equation for α is derived from phase continuity equation and reads:

$$\frac{\partial \alpha}{\partial t} + \nabla \cdot (\mathbf{u}\alpha) + \nabla \cdot (\mathbf{u}^r \alpha(1-\alpha)) = 0\,, \tag{9.6}$$

where the last term is responsible for interface compression as described in [10]. Novel formulation of compressive flux, $\mathbf{u}^r$ is given in next section.

9.2.3 Wave Modelling Using Relaxation Zones

In CFD simulations, it is necessary to introduce incoming waves which do not arise naturally from governing equations. Since it is practically difficult to impose

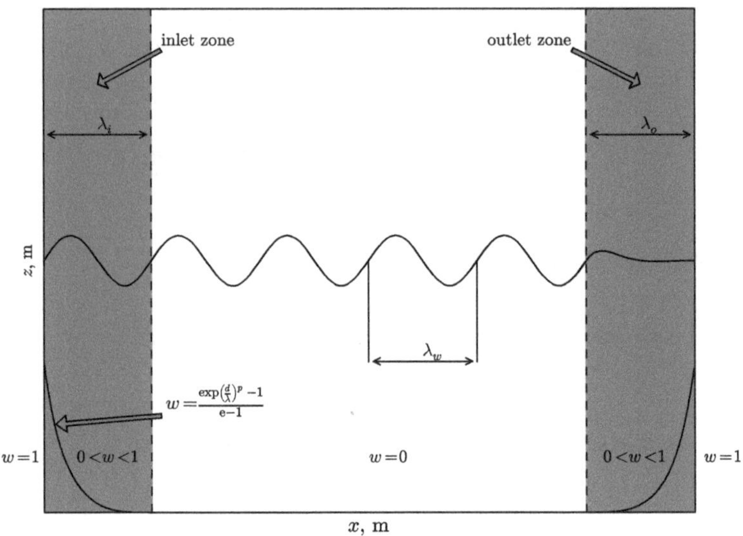

Fig. 9.1 Relaxation zones with weight field

wave conditions on boundaries that will correspond to the given sea state, we shall introduce the concept of relaxation zones [6]. In such zones, wave modelling is conducted with coupling of potential flow wave theories and CFD solutions. They are used to smooth out the transition between wave theory and CFD solution, as seen in Fig. 9.1. The inlet relaxation zone is used to generate incoming waves while the outlet relaxation zone to damp out outgoing waves without reflection, preventing the pollution of CFD results. The idea of relaxation zone is to volumetrically combine the governing equations, with solutions of simplified wave theory equations. Since wave theories are derived from the Navier–Stokes equations, their solution shall be consistent with the governing laws and thus be available for blending. It is however necessary to account for simplifications in wave theory solutions in a consistent manner through the implicit blending procedure. The procedure is explained with general transport equation for variable ϕ:

$$\frac{\partial(\rho\phi)}{\partial t} + \nabla \cdot (\rho \mathbf{u} \phi) - \nabla \cdot (\gamma_\phi \nabla \phi) - S_u + S_p \phi = \mathcal{T}(\phi) = 0\,, \tag{9.7}$$

where γ_ϕ is the diffusion coefficient, S_u the source term and $S_p\phi$ the linearized sink term. $\mathcal{T}(\phi)$ denotes the general transport operator. Now, consider that $\phi_t(\mathbf{x}, t)$ is the known, target solution obtained with different mathematical model (e.g. potential flow wave theory). Then, the following equation can be written on boundaries:

$$\phi = \phi_t \rightarrow \phi - \phi_t = \mathcal{R}(\phi) = 0\,. \tag{9.8}$$

In the above equation, $\mathcal{R}(\phi)$ is the relaxation zone operator. Weight field, w is used to blend two models represented by Eqs. (9.7) and (9.8). w is equal to 1 at the boundaries (inlet and outlet, see Fig. 9.1), forcing the target solution. Toward the domain interior, w changes smoothly to 0 in order to force the full CFD solution. Finally, a single equation can be written in terms of w and Eqs. (9.7) and (9.8):

$$(1-w)\mathcal{T}(\phi) + w\mathcal{R}(\phi) = 0\,, \tag{9.9}$$

This will force the target solution given by Eq. (9.8) where $w = 1$, and give the full CFD solution (9.7) where $w = 0$. w is most often chosen to be of the following form [6]:

$$w = \frac{\mathrm{e}^{\left(\frac{d}{\lambda}\right)^p} - 1}{\mathrm{e} - 1}\,, \tag{9.10}$$

where d represents the shortest distance to the boundary and λ the length of relaxation zone. p is the spatial exponent, usually set to 3.5. It should be noted that the length of the outlet relaxation zone should be approximately equal to the wave length in order to prevent any wave reflection. In this approach, velocity field and volume fraction are blended in this manner, while the dynamic pressure equation is derived from the continuity equation in the usual manner [7].

9.3 Numerical Model

Governing equations presented in Sect. 9.2 are implemented in OpenFOAM's `Naval Hydro` pack. Pressure-velocity coupling is resolved with a combined segregated algorithm that uses a few PISO correctors within each SIMPLE corrector. This treatment allows a number of successive solutions of VOF equation (9.6) in each time step in order to properly couple the density with velocity and pressure. Every equation is treated implicitly, including the VOF transport equation. Such procedure with implicit blending using relaxation zones allows stable simulations at maximum Courant–Friedrichs–Lewy (CFL) numbers higher than 1. The compressive flux needed for FVM discretization in Eq. (9.6) is defined as follows:

$$\mathbf{u}^r \bullet \mathbf{S} = c_\alpha \hat{n}_\Gamma \min\left(\frac{\mathrm{CFL_{ref}}\Delta x}{\Delta t}, \frac{\mathbf{u}\bullet\mathbf{S}}{|\mathbf{S}|}\right)\,, \tag{9.11}$$

where c_α represents the compression constant usually taken as 1, $\hat{n}_\Gamma$ is the free surface normal vector that has the magnitude of the surface area. $\mathrm{CFL_{ref}} = 0.5$ is the reference compression CFL number and Δx denotes the distance between cell centres encompassing the face. This formulation makes the compressive flux independent of the physical flux through the free surface.

Linear upwind scheme is used for momentum convection while the van Leer, Total Variation Diminishing (TVD) scheme [13] is used for volume fraction convection. For compressive convection term, special `vofCompression` scheme is used, which smoothly switches from central differencing in regions where $\alpha \approx 0.5$ to upwind differencing where $\alpha \approx 1$ or $\alpha \approx 0$. Hence, second order accuracy in space is achieved.

9.4 Test Cases

Two types of simulations are carried out with this model. The first simulation is used to validate the accuracy of the numerical simulation for regular waves. Wave loads on vertical cylinder are calculated and results are compared with experimental data from [1]. The second case shows the possibility of freak wave simulations.

9.4.1 Harmonic Wave Loads on Vertical Cylinder

The results of CFD simulations of wave loads on a vertical cylinder are presented for five different incoming waves. The experimental measurements are carried out in a tank 36.5 m long, 2.4 m wide and 1.5 m deep. The circular cylinder is vertically immersed in water to the depth of 0.27 m; it does not reach the bottom of the tank. Cylinder diameter is 89 mm, and it is placed at 13.7 m from the wave maker. At the opposite end of the tank from the wave marker, a sloping beach is placed for wave absorption. There is enough room between the cylinder and the beach for the force measurements to be carried out before the reflected waves reach the cylinder, polluting the results. In [1] maximum measured forces in the longitudinal direction are given. Maximum values are determined over ten wave periods for each wave. The measurements are carried out for wave frequencies of 1.43, 1.1, 1.0, 0.9 and 0.7 Hz. For each frequency ten wave slopes ranging from 0.06 to 0.24 are used, where $k\eta_a$ is the wave slope, k wave number and η wave amplitude according to first order Stokes wave theory. Five waves are selected for numerical simulation and their parameters are presented in Table 9.1.

Table 9.1 Incident wave parameters

Index N	Frequency f (hz)	Wave slope $k\eta_a$ (rad)	Wave number k (rad/m)	Wave height h (m)	Wave length λ (m)	Period T (s)
1	0.70	0.06	1.97	0.060	3.19	1.43
2	0.70	0.12	1.97	0.120	3.19	1.43
3	0.90	0.20	3.26	0.123	1.93	1.11
4	1.10	0.12	4.87	0.050	1.30	0.90
5	1.43	0.20	8.83	0.049	0.76	0.70

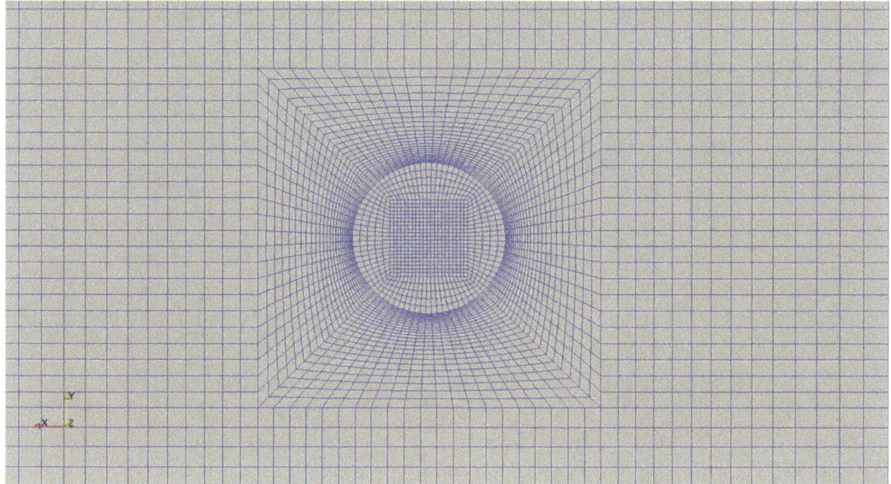

Fig. 9.2 Mesh geometry near the cylinder

The domain is reduced compared to the experimental setup to save computational time. Length of the domain is shortened to three wave lengths. The long tank is not needed since waves are absorbed completely in the outlet relaxation zone. Length of the inlet relaxation zone is set to half of the wave length, while the outlet relaxation zone is one wave length long. Smaller outlet relaxation zones proved to be reflective.

Figure 9.2 shows the finite volume mesh in horizontal plane near the cylinder. Mesh used for calculations consists of 1,728,490 hexahedral cells. Maximum cell aspect ratio is 1:47. Linear grading is used to refine the mesh around the cylinder. As a result, cells near the cylinder are 140 times smaller than those near the boundaries.

Wave forces in longitudinal direction are obtained for each wave from Table 9.1. Figure 9.3 shows a representative force signal for the simulation of wave indexed by number 3. Table 9.2 presents the comparison of the results obtained by numerical simulation and experimental results for each wave. For waves 1, 2 and 3, relative errors are within acceptable range. For higher frequencies the errors are larger. The accuracy of the calculation depends on the vertical mesh refinement, i.e. on the number of cells per wave height and on wave frequency. It can be seen in Table 9.2 that the error for wave 5 is significantly reduced when one uses finer mesh. For wave 4 it is still not clear why the mesh refinement did not produce better results compared to the experiment. This is the topic of ongoing investigation. It should be noted that finer meshes used for waves 4 and 5 are still relatively coarse mesh considering domain size, and it is believed that further refinement would improve the result. Furthermore, since waves 4 and 5 have smaller wave lengths (regarding fixed distance between wave maker and the cylinder in experimental setup), non-linear effects should be more expressed at the time of the impact. Better results could probably be obtained with high order wave theories, but this was not investigated in the present study.

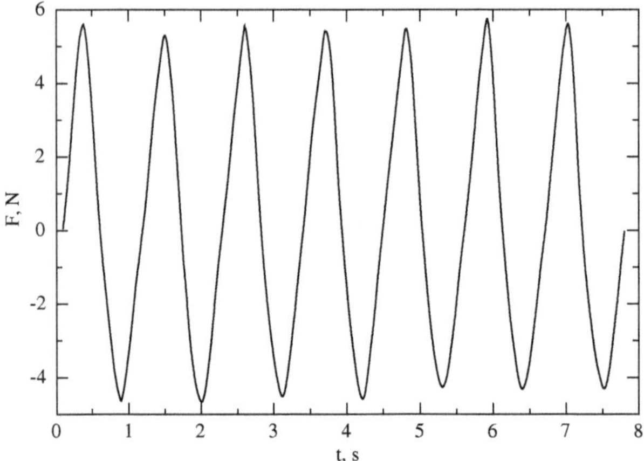

Fig. 9.3 Force signal for the third wave (Table 9.1)

Table 9.2 Comparison of CFD and experimental results

Wave index N	CFD results F_x (N)	Experimental results F_x (N)	Relative error Err (%)	Number of cells	Courant number Co	α Courant number αCo
1	1.778	1.80	1.22	1,728,490	6.0	3.00
2	4.790	5.00	4.20	1,728,490	6.0	3.00
3	5.573	5.70	2.23	1,728,490	2.0	1.50
4	2.390	2.80	14.64	1,728,490	1.5	0.75
4	2.361	2.80	15.68	2,805,810	1.5	0.75
5	2.650	3.08	13.96	1,728,490	2.0	1.50
5	2.854	3.08	7.34	2,629,410	2.0	1.50

9.4.2 Freak Wave Simulation

Freak wave, also known as rouge wave, is a phenomenon that is not fully understood [8]. According to the definition that is most widely accepted, a freak wave is a wave whose height exceeds the significant wave height of the current sea state by two times [8]. Freak wave has a low probability of occurrence, but it poses a great threat to offshore structures and ships, which is why new methods of calculating freak wave loads are getting more attention.

Harmonic wave focusing method is used to generate the freak wave. Harmonic waves used for focusing are components of Pierson–Moskowitz sea spectrum. This is done in order to obtain a more realistic freak wave, since sea spectrum describes realistic sea states. Free surface elevation is determined by superimposing individual harmonics. For this simulation, 30 wave components are used, guided by recommendation in [15]. Phase shifts of wave components are usually selected from a uniformly distributed random numbers. This ensures that the reproduced sea state is statistically equal to the sea state on which the wave energy spectrum is determined. However, to initialize a freak wave, phase angles must be set in a way that will ensure positive superposition of the wave components in order to achieve extreme wave height at desired time and position. Thus phase shifts are obtained by an optimization procedure. Freak wave in this study is obtained by linear superposition of wave components and is 0.255 m high, which is 2.12 times higher than selected significant wave height of 0.12 m.

Figure 9.4a depicts the freak wave profile in the simulation shortly before the focusing time of 2.66 s. Blue colour shows the part of the domain which contains water ($\alpha = 1$), while red colour shows the air in the domain ($\alpha = 0$). The thick horizontal white line in Fig. 9.4a is positioned at calm free surface level. Thinner white horizontal lines show positive and negative significant amplitudes, i.e. distance between them is the significant wave height. Black horizontal lines are positioned at the crest and trough of the freak wave. Height difference is obvious, and it agrees well with the calculated height. The white vertical line shows the position of the cylinders centreline. Figure 9.4b shows a perspective view of the simulation at the same instant of time as in Fig. 9.4a. Propagation of the freak wave caused further steepening. Figure 9.5a presents the freak wave impact on the circular

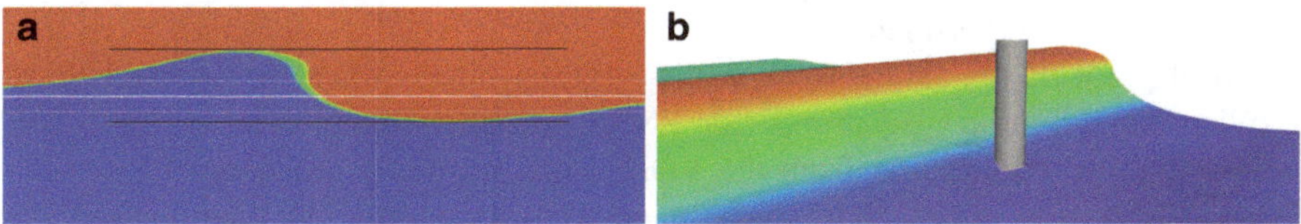

Fig. 9.4 Freak wave at the time instant before the wave impact. (**a**) Profile view. (**b**) Perspective view

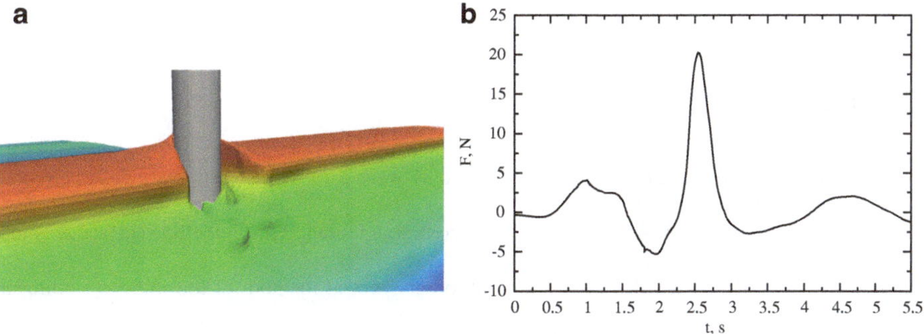

Fig. 9.5 Freak wave impact. (**a**) Perspective view at the impact. (**b**) Longitudinal force exerted by the freak wave

cylinder. It can be observed that the wave was close to vertical at the time of impact. This is in accordance with often encountered description of freak wave as a "wall of water". Figure 9.5b presents excessive force caused by the freak wave at the time of the impact.

9.5 Conclusion and Future Work

This chapter described a two-phase incompressible CFD flow model specialized for offshore applications. Coupling with potential flow wave theories is obtained with implicit relaxation zones in second-order accurate FVM.

Numerical simulations of two types of problems are carried out which have practical application in modern offshore industry. The first case regarding regular wave loads agrees well with experimental results. The second case presents the possibility of evaluating loads on offshore structures and ships exerted by the freak wave, which is hardly achievable using other conventional methods. The results show that numerical simulations in OpenFOAM may indeed be used in the design process of offshore structures. Considerable savings could be made by using CFD results to get more detailed insight in the governing physics, gradually leading to a lower number of experiments.

Focal point of the future work will be further development of Spectral Wave Explicit Navier–Stokes Equations (SWENSE) method implemented in Finite Volume framework. Transition from the VOF to the implicitly redistanced Level Set interface capturing approach has been made. This allows efficient coupling of CFD near the structure of interest with arbitrary potential flow solutions in far field. Furthermore, the group will try to implement the Higher Order Spectrum (HOS) method [3, 14] in order to improve the freak wave simulations.

References

1. Boo SY (2006) Measurements of higher harmonic wave forces on a vertical truncated circular cylinder. Ocean Eng 33:219–233
2. Brackbill JU, Kothe DB, Zemach C (1992) A continuum method for modelling surface tension. J Comput Phys 100:335–354
3. Dommermuth DG, Yue DKP (1987) A high-order spectral method for the study of nonlinear gravity waves. J Fluid Mech 184:267–288
4. Higuera P, Lara JL, Losada IJ (2013) Realistic wave generation and active wave absorption for Navier-Stokes models: application to OpenFoam®. Coast Eng 71:102–118
5. Higuera P, Lara JL, Losada IJ (2013) Simulating coastal engineering processes with OpenFoam®. Coast Eng 71:119–134
6. Jacobsen NG, Fuhrman DR, Fredsøe J (2012) A wave generation toolbox for the open-source CFD library: OpenFoam®. Int J Numer Methods Fluids 70(9):1073–1088
7. Jasak H (1996) Error analysis and estimation for the finite volume method with applications to fluid flows. PhD thesis, Imperial College of Science, Technology & Medicine

8. Kharif C, Pelinovksy E (2003) Physical mechanisms of the rogue wave phenomenon. Eur J Mech B/Fluids 22:603–634
9. Monroy C, Ducrozet G, Bonnefoy A, Babarit A, Ferrant P (2010) RANS simulations of a CALM buoy in regular and irregular seas using the SWENSE method. In: Proceedings of the twentieth (2010) international offshore and polar engineering conference, Beijing, 20–25 June 2010
10. Rusche H (2002) Computational fluid dynamics of dispersed two - phase flows at high phase fractions. PhD thesis, Imperial College of Science, Technology & Medicine
11. Ubbink O (1997) Numerical prediction of two fluid systems with sharp interfaces. PhD thesis, Imperial College of Science, Technology & Medicine
12. Ubbink O, Issa RI (1999) A method for capturing sharp fluid interfaces on arbitrary meshes. J Comput Phys 153:26–50
13. van Leer B (1977) Towards the ultimate conservative difference scheme. IV. A new approach to numerical convection. J Comput Phys 23:276–299
14. West BJ, Brueckner KA, Janda RS (1987) A new numerical method for surface hydrodynamics. J Geophys Res 92(C11):11803–11824
15. Zhao X (2010) Numerical simulation of extreme wave generation using VOF method. J Hydrodyn 22:466–477

Chapter 10
MLS-Based Selective Limiting for Shallow Waters Equations

J. Cernadas, X. Nogueira, and I. Colominas

Abstract Finite Volume Methods were successfully used in the last years to solve differential hyperbolic problems (Leveque, Finite-volume methods for hyperbolic problems. Cambridge University Press, Cambridge, 2005). Our research is focused here on the use of the Moving Least Squares (*MLS*) approximations for the development of a selective limiting technique to keep the accuracy of high-order methods in non-smooth flows. Following (Nogueira et al., Comput Methods Appl Mech Eng 199:2544–2558, 2010) we use the multiresolution properties of the *MLS* methodology and we define a shock-detection technique to act as a smoothness indicator. This sensor is used to detect shock waves present in the flow problem. The use of this technique combined with slope limiters improves the accuracy of the resulting TVD scheme. In this work we present the first results obtained with this technique applied to the resolution of the shallow waters equations with a high-order *FV-MLS* scheme (Cueto-Felgueroso et al., Comput Methods Appl Mech Eng 196:4712–4736, 2007). We present several 1D results and we compare them with those obtained with other high-order schemes.

10.1 Introduction

The conversion of tidal energy into electricity through the use of turbines is a promising mechanism to obtain electric energy [4]. The reason is that tides and sea flows are more predictable than other sources of renewable energy as, for example, wind. The *Shallow Waters Equations* are a mathematical model that allows us to predict shallow waters flows, and evaluate the available resource.

There exist several methods to solve these equations [12]. In this work we present our first results of a new technique that improves the results of a finite volume [8] numerical scheme.

J. Cernadas • X. Nogueira (✉) • I. Colominas
Civil Engineering School, University of A Coruña, A Coruña, Spain
e-mail: jesus.cernadas@udc.es; xnogueira@udc.es; icolominas@udc.es

E. Ferrer, A. Montlaur (eds.), *CFD for Wind and Tidal Offshore Turbines*,
Springer Tracts in Mechanical Engineering, DOI 10.1007/978-3-319-16202-7_10

In tidal turbine flows, there are some phenomenons that may produce shock waves (or more generally flow discontinuities). This is the case of hydraulic jumps which occur when water at high velocity discharges into a zone of lower velocity.

When we solve the shallow waters problem with a higher-order finite volume scheme [1, 2, 9, 10], it is difficult to keep the accuracy of the method if shock waves are present in the domain. Shock waves introduce non-smooth regions where the numerical scheme goes to order one. Here, we propose the use of *Moving Least Squares* (*MLS*) [10] for the development of a selective limiting technique to keep the accuracy of high-order methods in non-smooth shallow water flows. Using the multiresolution properties of the *MLS* approximations, we define a shock wave detection technique to act as a smoothness indicator and we combine it with the use of slope limiters, to develop a *selective limiting* method that allows us to keep the high order of the numerical scheme even under the presence of shock waves.

In next sections, we present the mathematical model of 1D shallow waters flows, a brief overview of *MLS* approximations, and the new *selective limiting* technique. Finally we present some results and conclusions.

10.2 Shallow Waters Equations

The 1D shallow waters equations can be written as

$$\mathbf{U}_t + \mathbf{F}(\mathbf{U})_x = \mathbf{S}(\mathbf{U}) , \tag{10.1}$$

with

$$\mathbf{U} = \begin{bmatrix} h \\ hu \end{bmatrix}, \quad \mathbf{F}(\mathbf{U}) = \begin{bmatrix} hu \\ hu^2 + \frac{1}{2}gh^2 \end{bmatrix}, \quad \mathbf{S}(\mathbf{U}) = \begin{bmatrix} 0 \\ -ghb_x \end{bmatrix}, \tag{10.2}$$

where u, v, w are the components of fluid velocity and p is the pressure of the fluid. We neglect viscous effects but we retain body forces via a source term vector. We assume that the density of the fluid (ρ) is constant and we consider the fluid under the action of gravity through the source term vector $\mathbf{g} = (0, 0, -g)$, where $g = 9.8\,\mathrm{m/s^2}$ is the acceleration due to gravity. Making the assumption that the bottom is a boundary defined by a function $z = b(x, y)$, we define the *free surface* of the flow as $z = s(x, y, t) \equiv b(x, y) + h(x, y, t)$, where $h(x, y, t)$ is the depth of water.

10.3 *MLS*-Based Selective Limiting

In this work, we develop an *MLS*-Based selective limiting technique to improve the accuracy of finite volume methods with slope limiters for the *1D Shallow Water Equations*. The use of slope limiters is an usual technique to build TVD methods

[7]. It is commonly used with second-order finite volume schemes. In order to keep the stability of the numerical scheme when the flow is non-smooth, a slope limiter limits the Taylor reconstruction of a function $\mathbf{U}(\mathbf{x})$ in the framework of high order finite volume schemes as follows:

$$\mathbf{U}(\mathbf{x}) = \mathbf{U_I} + \chi_I \nabla \mathbf{U_I} \cdot (\mathbf{x} - \mathbf{x_I}) \,, \tag{10.3}$$

where χ_I is a parameter between 0 and 1, computed with some slope-limiter algorithm. In our particular case, we limit the reconstruction of two variables: the depth of water h and the x component of velocity, u. Thus, the limited Taylor reconstruction for these two variables is, respectively

$$\begin{aligned} h(x) &= h_I + \chi_{hI} \nabla h_I (x - x_I) \\ u(x) &= u_I + \chi_{uI} \nabla u_I (x - x_I) \,. \end{aligned} \tag{10.4}$$

To obtain χ_{hI} and χ_{uI}, we use the limiter of van-Albada [12]. The value of a van-Albada slope limiter χ_{hI} depends on a parameter r_i. The expression for this slope limiter is

$$\chi_i = \begin{cases} 0 & \text{if } r_i \leq 0 \\ \min\left\{ \frac{r_i(1+r_i)}{1+r_i^2} \,,\, \frac{4}{(1-c)(1+r_i)} \right\} & \text{if } r_i \geq 0 \end{cases} , \qquad i = hI,\, uI \,. \tag{10.5}$$

c is the Courant number [11], given by $\frac{\lambda}{\Delta x / \Delta t}$, where λ is the wave propagation speed, Δx is the cell width, and Δt is the size of the time step used in the computation.

The value of r_i for a variable i can be obtained as

$$r_i = \frac{i_I - i_{I-1}}{i_{I+1} - i_I} \,. \tag{10.6}$$

The use of slope limiters can improve the results of a numerical scheme when the solution is not smooth, but they also present some drawbacks: for example, they can avoid the convergence of the numerical scheme, moreover they may be active in cells where the flow is smooth and where the limiter should not be active. Other drawback is the fact that they can give anomalies in physical parameters as the entropy [10] and finally the straightforward application to higher-order schemes is not obvious. In order to alleviate these drawbacks, we propose the use of the multiresolution properties of MLS to develop a selective limiting technique. In the next subsections, we present the basis of *MLS* approximations, and the use of *MLS* properties to act as a filter and to develop the shock detector.

10.3.1 MLS Formulation

The basic idea for the *MLS* approaching method is to reconstruct the value of a certain function $u(\mathbf{x})$ at a point $\mathbf{x}$ by using a weighted Least Squares approximation in a compact support ($\Omega_{\mathbf{x}}$) on the vicinity of $\mathbf{x}$. For the sake of brevity, here we only show the final form of the approximation (we refer the interested reader to [6] for more details) that can be written as

$$u(\mathbf{x}) \approx \hat{u}(\mathbf{x}) = \sum_{j=1}^{n_{\mathbf{x}}} N_j(\mathbf{x}) u_j = \mathbf{p}(\mathbf{x})^T \mathbf{M}^{-1}(\mathbf{x}) \mathbf{P}_{\Omega_{\mathbf{x}}} \mathbf{W}(\mathbf{x}) \mathbf{u}_{\Omega_{\mathbf{x}}} . \tag{10.7}$$

In this expression, $N_j(\mathbf{x})$ are the *MLS* shape functions and $n_{\mathbf{x}}$ is the number of neighboring cells of the stencil of the cell where the approximation is performed. $\mathbf{u}_{\Omega_{\mathbf{x}}}$ is the vector that contains the known values of $u(\mathbf{x})$ at $\Omega_{\mathbf{x}}$. $\mathbf{p}(\mathbf{x})$ is a basis of m functions and $\mathbf{P}_{\Omega_{\mathbf{x}}}$ is defined as a matrix where the basis functions are evaluated at each point of the stencil. $\mathbf{W}(\mathbf{x})$ is a smoothing function called *kernel* (it acts as a weight function in the *MLS* reconstruction of a variable). Finally, $\mathbf{M}(\mathbf{x})$ is the moment matrix, defined by $\mathbf{M}(\mathbf{x}) = \mathbf{P}_{\Omega_{\mathbf{x}}} \mathbf{W}(\mathbf{x}) \mathbf{P}_{\Omega_{\mathbf{x}}}^T$ and obtained by a minimization of an error functional [2].

10.3.1.1 The Basis of Functions p(x)

The basis of functions $\mathbf{p}(\mathbf{x})$ is a key parameter in the *MLS* reconstruction. It has been shown [5] that for a polynomial basis of order r, the nominal order of accuracy to calculate the s-order derivatives is $(r-s+1)$. With *MLS* we can approximate exactly every lineal combination of functions that are included in the basis of functions, so it is important to choose them carefully. The most used basis of functions are polynomial basis, and here, we use a 1D cubic basis of polynomial functions $\mathbf{p}(\mathbf{x}) = (1, x, x^2, x^3)$.

10.3.1.2 The Cloud of Points

In this work, we use the FV-*MLS* [1, 2, 9, 10] numerical scheme. In this method it is necessary to define some points (neighbor points) that influence the accuracy of the reconstruction. In our particular case, we use the centroids of neighbor cells as neighbor points. This *cloud of points* allow us to perform the *MLS* reconstruction.

Before the reconstruction, we need to define the stencil, that is, to define the points that will be part of the cloud of points. In our particular case, we consider that the stencil does not change with each time step, so we only need to define it once at the beginning of the computations. The stencil has to meet some requirements to ensure that the method works correctly. If we have a basis with m functions, we need a stencil with more than m cells. Taking a number of cells greater than

m, we guarantee that the moment matrix can be inverted. Here, we use a 1D cubic basis, so we choose a stencil of five cells, centered in the point where we make the reconstruction of the variables [9].

10.3.1.3 The Kernel Functions

Kernel functions weight the relative importance of the neighbor points in the reconstruction. There exist many kinds of kernel functions (exponential functions, spline functions, etc.). In this work, we use the following cubic spline kernel function

$$W(s) = \begin{cases} 1 - \frac{3}{2}s^2 + \frac{3}{4}s^3 & s \leq 1 \\ \frac{1}{4}(2-s)^3 & 1 < s \leq 2 \\ 0 & s > 2 \end{cases}, \tag{10.8}$$

where $s = |(x - x_I)/(h_s)|$.

The smoothing length ($h_s = k \max |x - x_I|$) depends on a parameter k called shape parameter, and it limits the size of the compact support where we define the kernel functions. It is important to choose an appropriate value of h_s because the kernel function determines the numerical properties of the approximation [9] and of the *MLS* filter presented in the next section. Figure 10.1 shows an example of a cubic kernel for some values of the shape parameter.

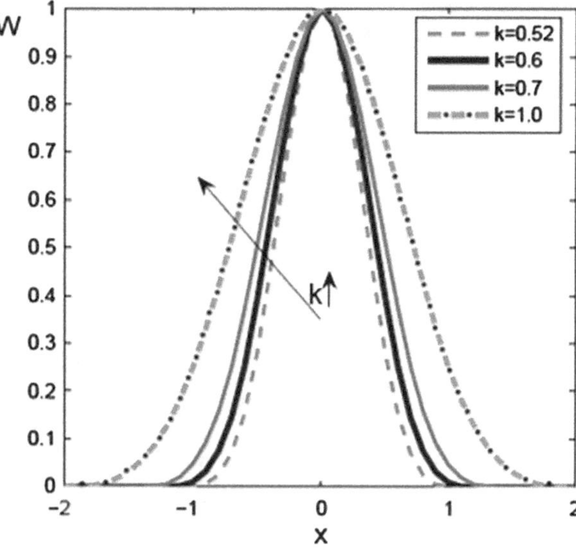

Fig. 10.1 Cubic kernel for different values of the shape factor k that defines the smoothing length h_s

10.3.2 MLS-Based Filters

It is possible to see the *MLS* approximation of a variable as a low-pass filtering. Thus, for a given variable Φ, we can write

$$\overline{\Phi_I} = \sum_{j=1}^{n} N_j(\mathbf{x})\Phi_j \,, \tag{10.9}$$

where n is the number of neighbors of the stencil of the cell I. The bar indicates a filtered variable. The filter properties can be analyzed through the study of the transfer function. The transfer function associated with the filter given by Eq. (10.9) is

$$\hat{G}(\omega) = \sum_{j=1}^{n} N_j(\mathbf{x})e^{i\omega(x_j - x_I)} \,, \tag{10.10}$$

where ω is the wavenumber. $\hat{G}(\omega)$ depends on the stencil, the basis of functions and the kernel functions. For a cubic kernel, if we increase the value of the shape parameter k we also increase the range of frequencies that we filter. Figure 10.2 shows the transfer function for a cubic kernel [10].

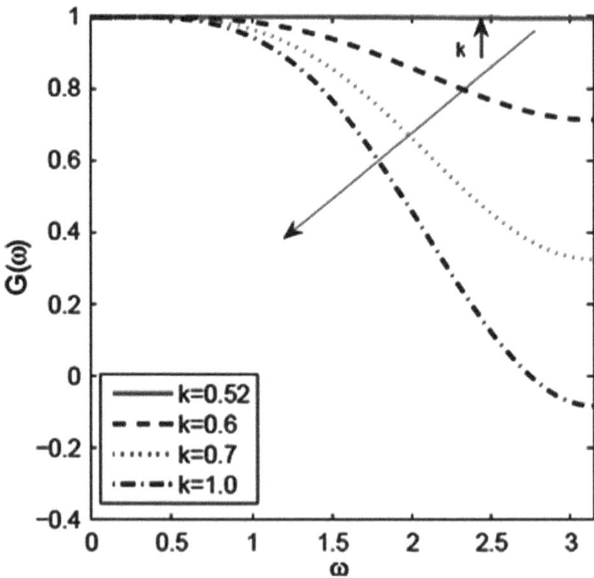

Fig. 10.2 1D transfer function for a cubic kernel with different values of k

10.3.3 MLS-Based Shock Detection Method

As commented previously, the use of higher-order schemes with slope limiters introduces a number of drawbacks. For this reason, we have developed a methodology to improve the limitation technique. This technique determines the points where the limiter algorithm is active. The limiter should be activated in the areas where the solution is non-smooth. We use the multiresolution properties of *MLS* to develop a sensor able to detect the points where the limiter must be active. We recall that *MLS* approximations are related with wavelets [3]. A wavelet is defined by a wavelet function and a scale function, and both of them can be defined using *MLS* approximations. Thus, to obtain the scale function, we consider the *MLS* approximation

$$h(\mathbf{x}) \approx h^{h_s}(\mathbf{x}) = \sum_{j=1}^{n_I} h_j N_j{}^{h_s}(\mathbf{x}) \,, \tag{10.11}$$

where h is the depth of water. The approximated solution $h^{h_s}(\mathbf{x})$ keeps all the resolutions and properties of the solution $h(\mathbf{x})$, up to scale h_s (given by the smoothing length). We can see the *MLS* shape functions as the h_s-scale function of a wavelet, and h_s is the scale parameter. A smaller value of h_s is related to a higher level of resolution.

We consider a function $h(\mathbf{x})$, and we define two sets of *MLS* shape functions $\mathbf{N}^{h_s}(\mathbf{x})$ and $\mathbf{N}^{2h_s}(\mathbf{x})$, with two different smoothing lengths: h_s and $2h_s$, respectively. Thus, we obtain an h_s-approximation and a $2h_s$-approximation (high and low level of resolution, respectively), of $h(\mathbf{x})$

$$\mathbf{h}^{h_s}(\mathbf{x}) = \sum_{j=1}^{n_I} \mathbf{h}_j N_j{}^{h_s}(\mathbf{x}) \,, \quad \mathbf{h}^{2h_s}(\mathbf{x}) = \sum_{j=1}^{n_I} \mathbf{h}_j N_j{}^{2h_s}(\mathbf{x}) \,. \tag{10.12}$$

Note that $\mathbf{h}_j$ are the values of the depth of water $h(\mathbf{x})$ at the neighboring cells. The wavelet function is then defined as

$$\boldsymbol{\phi}(\mathbf{x}) = \mathbf{N}^{h_s}(\mathbf{x}) - \mathbf{N}^{2h_s}(\mathbf{x}) \,. \tag{10.13}$$

We can express the h_s solution as the sum of its low scale and high scale complementary parts, as

$$\boldsymbol{\psi}(\mathbf{x}) = \mathbf{h}^{h_s}(\mathbf{x}) - \mathbf{h}^{2h_s}(\mathbf{x}) = \sum_{j=1}^{n_I} \mathbf{h}_j (N_j{}^{h_s}(\mathbf{x}) - N_j{}^{2h_s}(\mathbf{x})) = \sum_{j=1}^{n_I} \mathbf{h}_j \boldsymbol{\phi}_j(\mathbf{x}) \,. \tag{10.14}$$

The function $\psi(\mathbf{x})$ acts as a smoothness indicator of $h(\mathbf{x})$. In a cell the limiter is activated if the value of $\psi(\mathbf{x})$ is greater than a threshold value. Thus, the function $\psi(\mathbf{x})$ allows us to detect the presence of shock waves. Now, we have to define the threshold value. We use the depth of water as a reference variable, and we consider the gradient of h. We define the threshold value as

$$T_v = C_{\text{sc}} \left| \nabla h \right| , \tag{10.15}$$

where C_{sc} is a parameter. In our numerical experiments, the value $C_{\text{sc}} = 0.8$ gave us the best results in terms of robustness and accuracy. Thus, the slope limiter will be active when the following condition is verified

$$|\psi_h| = \left| \sum_{j=1}^{n_I} h_j \left(N_j{}^{h_s}(\mathbf{x}) - N_j{}^{2h_s}(\mathbf{x}) \right) \right| > T_v , \tag{10.16}$$

where h_j are the values of the depth of water at the neighboring points. If $C_{\text{sc}} = 0$, there is no selective limiting and the usual limitation algorithm is used in the whole domain. Note that the selective limiter switches on or off the whole limiter algorithm, but it is the original limiter who decides if the gradient is limited or not.

10.4 Numerical Results

In this section we present the results of three numerical tests. All of them can be considered as a simplification of a dam-break problem in a channel: after the initial instant, the dam disappears (breaks) and the water flow starts. We model the dam as an initial discontinuity between two masses of water. We place the discontinuity at the x_0 coordinate (in meters). In Table 10.1, we will show the parameters that define the tests we have included. The parameters h_L and h_R are the values of the depth of water (in meters) to the left and right side of the dam, respectively, in the initial time (the boundary values of the depth keep constant during all the test). In the same way, we define the values of the initial velocity (in m/s), denoted by u_L and u_R, respectively. The parameter t_{out} is the output time of the results, in seconds. We also define the batimetry (bottom) by $z = z(x)$.

Table 10.1 Data for three numerical tests

Test	$z(x)$	h_L (m)	u_L (m/s)	h_R (m)	u_R (m/s)	x_0 (m)	t_{out} (s)
1	Flat bottom	10	0	2	0	500	7
2	$0.8 \sin\left(\frac{\pi x}{1000}\right)$	10	0	1	0	500	7
3	$\sin\left(\frac{\pi x}{1000}\right)$	10	0	1	0	500	7

The length of the channel is 1,000 m

Figures 10.3, 10.4 and 10.5 present the results of these tests. We compare the solutions given by a second order scheme without using selective limiting (*O2*) and the numerical scheme that uses selective limiting (*O2 LS*) with the *Reference* solution, obtained by using a very fine mesh. In each pair of images, the left one shows the comparison of the solutions in a general view. For the case of selective limiting we mark the cells where the slope-limiter algorithm is active. In the image on the right, we show a closer view of the regions where we have found significant differences between both methods. In tests 1 and 2 we use a 100-element mesh (*100e*), and in test 3 we use both mesh sizes (*100e* and *200e*).

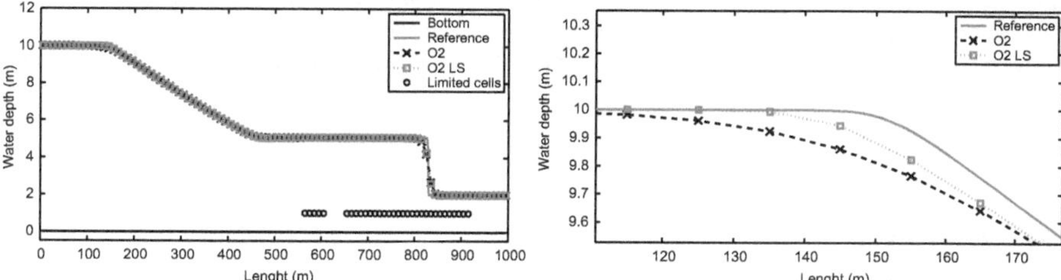

Fig. 10.3 Results for Test 1. Depth of water with (*square*) and without (*cross*) selective limiting. *Left figure* shows a general view and *right figure* shows a detailed view. On the *left figure*, cells where the slope-limiter algorithm is active are marked with *circles*

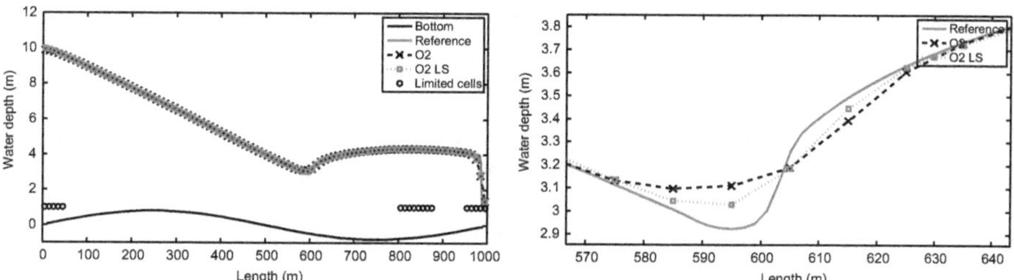

Fig. 10.4 Results for Test 2. Depth of water with (*square*) and without (*cross*) selective limiting. *Left figure* shows a general view and *right figure* shows a detailed view. On the *left figure*, cells where the slope-limiter algorithm is active are marked with *circles*

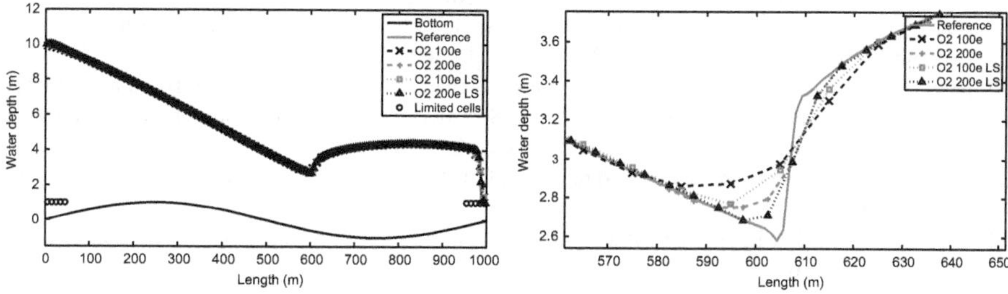

Fig. 10.5 Results for Test 3. Depth of water with and without selective limiting, for different mesh sizes. *Left figure* shows a general view and *right figure* shows a detailed view. On the *left figure*, cells where the slope-limiter algorithm is active are marked with *circles*

10.5 Conclusions

We have presented a new technique to improve the results obtained for a second-order finite volume method in the resolution of the shallow water equations.

An MLS-based sensor has been developed using the multiresolution properties of MLS approximations. It can be applied to both structured and unstructured grids and it improves the results of the numerical scheme without increasing the number of elements in the mesh.

Acknowledgements This work has been partially supported by the *Ministerio de Ciencia e Innovación* (#DPI2010-16496) and the *Ministerio de Economía y Competitividad* (#DPI2012-33622) of the Spanish Government and by the *Consellería de Cultura, Educación e Ordenación Universitaria* of the *Xunta de Galicia* (grant # GRC2014/039) cofinanced with FEDER funds, and the *Universidade da Coruña.*

References

1. Cueto-Felgueroso L, Colominas I, Fe J, Navarrina F, Casteleiro M (2006) High order finite volume schemes on unstructured grids using Moving Least Squares reconstruction. Application to shallow water dynamics. Int J Numer Methods Eng 65:295–331
2. Cueto-Felgueroso L, Colominas I, Nogueira X, Navarrina F, Casteleiro M (2007) Finite volume solvers and Moving Least-Squares approximations for the compressible Navier-Stokes equations on unstructured grids. Comput Methods Appl Mech Eng 196:4712–4736
3. Daubechies I (1992) Ten lectures on wavelets. Society for Industrial and Applied Mathematics, Philadelphia
4. Douglas CA, Harrison GP, Chick JP (2008) Life cycle assessment of the Seagen marine current turbine. Proc Inst Mech Eng M J Eng Marit Environ 222(1):1–12
5. Gossler A (2001) Moving least-squares: a numerical differentiation method for irregularly spaced calculation points. SANDIA Report, SAND2001-1669
6. Lancaster P, Salkauskas K (1981) Surfaces generated by moving least squares methods. Math Comput 155(37):141–158
7. Leveque RJ (1990) Numerical methods for conservation laws. Birkhauser, Basel
8. Leveque RJ (2005) Finite-volume methods for hyperbolic problems. Cambridge University Press, Cambridge
9. Nogueira X, Cueto-Felgueroso L, Colominas I, Khelladi S (2010) On the simulation of wave propagation with a higher order finite volume scheme based in Reproducing Kernel methods. Comput Methods Appl Mech Eng 199(155):1471–1490
10. Nogueira X, Cueto-Felgueroso L, Colominas I, Navarrina F, Casteleiro M (2010) A new shock-capturing technique based on Moving Least Squares for higher-order numerical schemes on unstructured grids. Comput Methods Appl Mech Eng 199:2544–2558
11. Toro EF (1999) Riemann solvers and numerical methods for fluid dynamics, 2nd edn. Springer, Heidelberg
12. Toro EF (2001) Shock-capturing methods for free-surface shallow flows. Manchester Metropolitan University

Chapter 11
A Comparison of Panel Method and RANS Calculations for a Horizontal Axis Marine Current Turbine

J. Baltazar and J.A.C. Falcão de Campos

Abstract In this work, a comparison between results of a panel method and a RANS solver is made for a horizontal axis marine current turbine in uniform inflow conditions. The panel method calculations were made with panel code PROPAN (Baltazar, On the modelling of the potential flow about wings and marine propellers using a boundary element method. Ph.D. thesis, Instituto Superior Técnico, 2008). A vortex pitch wake alignment model is considered for the blade wake. The RANS calculations were carried out with RANS code ReFRESCO (http://www.marin.nl/web/Facilities-Tools/CFD/ReFRESCO.htm). A comparison of the blade pressure distributions, wake geometry and thrust and power coefficients is made. A reasonable to good agreement of the blade distribution and turbine forces is seen between the codes. Comparison of the numerical calculations with experimental performance data available in the literature is also presented. In general, the trend of the thrust and power coefficients is well captured by the numerical methods.

11.1 Introduction

There has been a growing interest in the utilisation of hydrokinetic energy conversion systems for generation of electricity. One of the most promising systems is based on the use of horizontal axis turbines, also known as marine current turbines (MCTs) for hydrokinetic energy extraction [11]. The ability to predict the hydrodynamic performance of an MCT is essential for the design and analysis of such devices.

Reynolds-Averaged Navier–Stokes (RANS) methods are currently being applied for the analysis of the viscous flow around MCTs. For example, Lawson et al. [14] and Otto et al. [17] have used RANS codes to simulate the hydrodynamics of horizontal axis MCTs. Although RANS methods are well-established tools

J. Baltazar (✉) • J.A.C. Falcão de Campos
Marine Environment and Technology Center, Instituto Superior Técnico, Universidade de Lisboa, Lisboa, Portugal
e-mail: joao.baltazar@tecnico.ulisboa.pt; falcao.campos@tecnico.ulisboa.pt

E. Ferrer, A. Montlaur (eds.), *CFD for Wind and Tidal Offshore Turbines*,
Springer Tracts in Mechanical Engineering, DOI 10.1007/978-3-319-16202-7_11

that are able to accurately simulate the hydrodynamic turbine performance, the computational effort is still reasonably high due to the need of large discretisation levels of the computational domain.

On the other hand, the analysis of the potential flow around MCTs may be carried out with a boundary element method, also known as panel method. This method may be adequate and cost-effective for use in the design of such systems, when the flow is not dominated by viscous effects. This method has been widely used in the hydrodynamic analysis of marine propellers and there is an extensive literature on this field [12]. The panel method was first applied to a horizontal axis MCT in steady flow [4, 6] and in unsteady flow conditions [5]. The method was also applied for the prediction of sheet cavitation on both suction and pressure sides of the turbine blades [7].

The purpose of the present work is to make a comparison between results obtained by panel method PROPAN [3, 6] with RANS code ReFRESCO [18, 19] in order to obtain a better insight on the importance of the viscous effects and on the limitations of the inviscid flow model for an MCT. The comparison is carried out for a model scale turbine rotor tested by Bahaj et al. and published in [2]. The paper is organised as follows: A description of the numerical methods is given in Sect. 11.2; the comparison of the panel method with the RANS solver and experimental data is presented in Sect. 11.3; in Sect. 11.4, the main conclusions are drawn.

11.2 Numerical Methods

Let us first consider the turbine rotor geometry and the coordinate reference system for description of the problem. Next, the panel code PROPAN and the RANS code ReFRESCO are presented.

11.2.1 Coordinate System

Consider the rotor of a horizontal axis turbine with radius R placed in a fluid stream and rotating with constant angular speed Ω around its axis. The turbine rotor is made of K blades symmetrically distributed around an axisymmetric hub. We assume the fluid stream at upstream of the rotor to be uniform with velocity U in the direction of the rotor axis. The fluid is assumed to be incompressible with density ρ. Figure 11.1 shows the coordinate system used to describe the geometry and the fluid flow around the turbine rotor.

We introduce a Cartesian coordinate system (x, y, z) rotating with the turbine rotor, with the x axis coinciding with the turbine axis and the positive direction pointing downstream. The y axis coincides with the reference line of one of the blades and the z axis completes a right-hand system. The unit vectors in this system are denoted by $(\mathbf{e}_x, \mathbf{e}_y, \mathbf{e}_z)$. We use a cylindrical coordinate system (x, r, θ), with

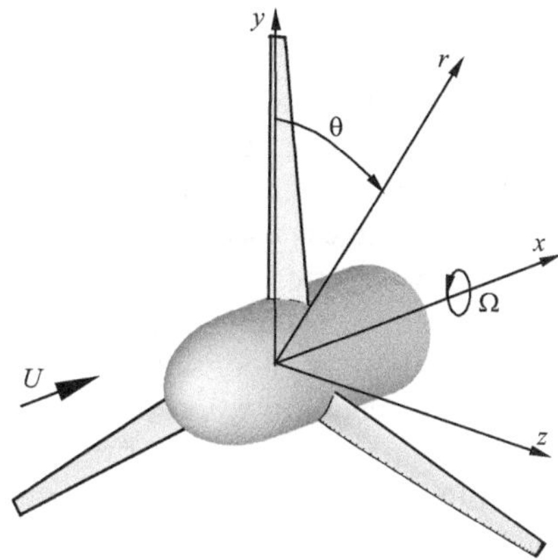

Fig. 11.1 Rotor coordinate reference systems

unit vectors $(\mathbf{e}_x, \mathbf{e}_r, \mathbf{e}_\theta)$, related to the Cartesian system by the transformation

$$y = r\cos\theta, \quad z = r\sin\theta. \tag{11.1}$$

The undisturbed onset velocity in the rotating frame is

$$\mathbf{U}_\infty = U\mathbf{e}_x + \Omega r\mathbf{e}_\theta. \tag{11.2}$$

11.2.2 Panel Code PROPAN

PROPAN is an IST (Instituto Superior Técnico) in-house panel code which implements a low-order potential-based panel method for the calculation of the incompressible inviscid potential flow around MCTs [3, 6].

Applying Green's second identity, assuming for the interior region to the blade surface S_B and hub surface S_H, the inner potential $\bar{\phi} = 0$, we obtain the integral representation of the perturbation potential ϕ at a point p on the blade or hub surfaces,

$$2\pi\phi(p) = \iint\limits_{S_B \cup S_H} \left[G\frac{\partial\phi}{\partial n_q} - \phi(q)\frac{\partial G}{\partial n_q} \right] dS - \iint\limits_{S_W} \Delta\phi(q)\frac{\partial G}{\partial n_q} dS, \quad p \in S_B \cup S_H \tag{11.3}$$

where $G(p,q) = -1/R(p,q)$, $R(p,q)$ is the distance between the field point p and the point q on the domain boundary $S_B \cup S_H \cup S_W$ and S_W is the wake surface. With the $\partial\phi/\partial n_q$ on the surfaces S_B and S_H known from the Neumann boundary condition on these rigid surfaces,

$$\frac{\partial\phi}{\partial n} \equiv \mathbf{n}\cdot\nabla\phi = -\mathbf{n}\cdot\mathbf{U}_\infty \quad \text{on} \quad S_B \cup S_H, \tag{11.4}$$

Eq. (11.3) is a Fredholm integral equation of the second kind in the dipole distribution $\mu(q) = -\phi(q)$ on the surfaces S_B and S_H. The Kutta condition yields the additional relationship between the dipole strength $\Delta\phi(q)$ on the wake surfaces S_W and the surface dipole strength at the blade trailing edges. An iterative pressure Kutta condition, which imposes equal pressure on both sides of the blade surface at the trailing-edge is used.

For the numerical solution of Eq. (11.3), we discretise the blade surfaces S_B, the hub surface S_H and the wake surfaces S_W in bi-linear quadrilateral elements which are defined by four points on the surface. The integral equation, Eq. (11.3), is solved by the collocation method with the element centre point as collocation point. We assume a constant strength of the dipole and source distributions on each element. The influence coefficients are determined analytically using the formulations given in [16].

A vortex pitch wake alignment model is considered for the blade wake. In this wake model the pitch of the vortex lines is aligned with the local fluid velocity, while the radial position of the vortex lines is fully prescribed. The description of the wake alignment model can be found in [8].

Corrections due to viscous effects to the inviscid turbine thrust and power calculated with the panel method are applied. The viscous forces on the turbine blades are calculated using the concept of section lift and drag force that can be derived from two-dimensional lift and drag data. The method is described in [6].

11.2.3 RANS Code ReFRESCO

ReFRESCO is a MARIN (Maritime Research Institute Netherlands) in-house viscous flow CFD code [18, 19]. It solves the multiphase unsteady incompressible RANS equations, complemented with turbulence models, cavitation models and volume fraction transport equations for different phases. The equations are discretised using a finite-volume approach with cell-centred collocation variables. A strong-conservation form and a pressure-correction equation based on the SIMPLE algorithm is used to ensure mass conservation [13]. The implementation is face-based, which permits grids with elements consisting of an arbitrary number of faces (hexahedrals, tetrahedrals, prisms, pyramids, etc.), and h-refinement (hanging nodes). The code is parallelised using MPI and sub-domain decomposition, and runs on Linux workstations and HPC clusters. ReFRESCO is currently being developed, verified and validated at MARIN (in the Netherlands) [9] in collaboration with IST (in Portugal), USP-TPN (University of Sao Paulo, Brasil), TUDelft (Technical University of Delft, the Netherlands), RuG (University of Groningen, the Netherlands) and recently at UoS (University of Southampton, UK).

For a turbine with uniform inflow conditions the equations can be solved using a so-called absolute formulation. This means that the velocity vector $\mathbf{V}$ is defined in the absolute or inertial earth-fixed reference frame, with the equations being solved in the body-fixed reference frame which is rotating with velocity Ω. This allows to perform steady simulations for this type of application. For all the calculations presented in this paper, the $\kappa - \omega$ SST 2-equation eddy-viscosity model proposed by Menter [15] is used. A second convection scheme (QUICK) is used for the momentum equations. A fine boundary layer resolution is applied to obtain $y^+ \sim 1$. However, in the present calculations a $y^+ > 1$ near the turbine tip is obtained and wall functions are used in that region.

11.3 Results

Results are presented for a model scale MCT tested by the Sustainable Energy Research Group of the University of Southampton [2]. The comparison of the calculations between the panel method and the RANS solver is shown for one tip-speed-ratio (TSR $= \Omega R/U$). Finally, the predicted thrust and torque coefficients by both methods are compared with the experiments for a range of TSRs.

11.3.1 General

The turbine rotor described in Bahaj et al. [2] is considered, where a considerable set of experimental data for an 800 mm diameter model in straight and yawed inflow conditions is available. The turbine is a three-bladed turbine with NACA 63-812, 63-815, 63-818, 63-821 and 63-824 sections. The standard geometry has a pitch angle at the blade root equal to 15°, corresponding to 0° pitch setting at the tip. In the present work, a 10° pitch setting angle has been considered. The detailed geometry of the MCT is given in Bahaj et al. [2].

Figure 11.2 shows the grids used for the calculations with panel code PROPAN and RANS code ReFRESCO. The discretisations of the PROPAN grid are: 60×31 on each blade, 270×30 on each blade wake and 80×48 on the hub. In the wake alignment model, the wake geometries were obtained after five iterations using 90 time steps per revolution which leads to an angular step of 4°. This discretisation provides a reasonably converged potential flow solution [6]. The viscous corrections to the blade forces derived from two-dimensional section lift and drag obtained from XFOIL computations published in [1].

The viscous calculations were also carried out for the MCT with RANS code ReFRESCO using a grid with 32 million cells. The grid was made with the commercial grid generation tool HEXPRESS™ [10], which uses an unstructured grid approach. A cylindrical domain is considered for the computational domain, where the inlet, outlet and outer boundaries are located 8 turbine rotor diameters from the origin of the coordinate system. At the inlet an uniform velocity and low turbulence level is prescribed, at the outlet an outflow condition is used and at the outer boundary a constant pressure is set. This discretisation level, domain size and boundary conditions were selected based on the numerical studies performed

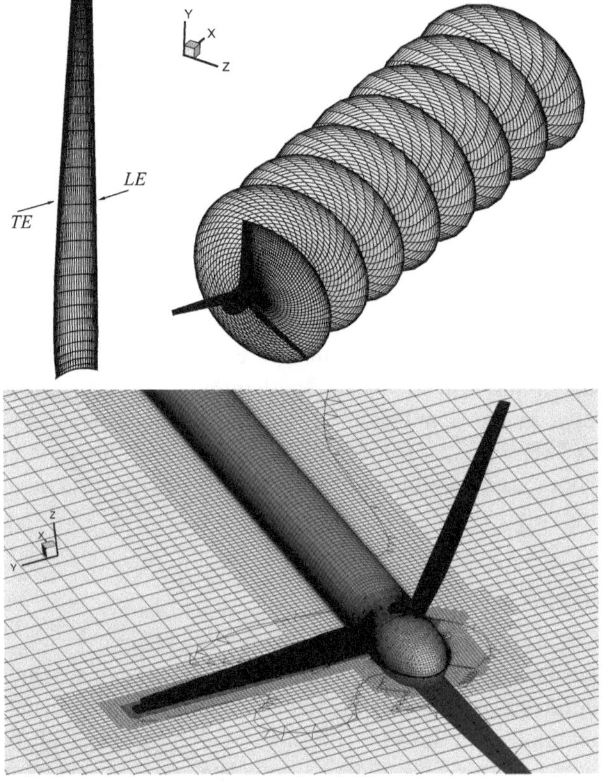

Fig. 11.2 Overview of the grids used for the inviscid calculations with PROPAN (*top*) and RANS calculations with ReFRESCO (*bottom*)

by Otto et al. [17]. The MCT was tested for a range of TSRs between 3.56 and 8.40 corresponding to Reynolds numbers from 1.29×10^5 to 2.86×10^5, where the Reynolds number is defined based on the turbine blade chord length at $0.7R$ ($c_{0.7R}$) and the resulting onset velocity at that radius:

$$\mathrm{Re} = \frac{c_{0.7R}\sqrt{U^2 + (n\pi 0.7D)^2}}{\nu}, \tag{11.5}$$

and $D = 2R$ is the propeller diameter, $n = \Omega/2\pi$ the rate of revolution and ν the fluid kinematic viscosity. Other useful quantities are the vorticity $\boldsymbol{\omega} = \nabla \times \mathbf{V}$ and the pressure coefficient, which can be defined based on the undisturbed onset velocity U_∞, Eq. (11.2), or based on the rotational speed nD:

$$C_p = \frac{p - p_\infty}{1/2\rho U_\infty^2}, \quad C_{pn} = \frac{p - p_\infty}{1/2\rho n^2 D^2}. \tag{11.6}$$

The non-dimensional thrust[1] and power of the MCT are given by the thrust coefficient C_T and the power coefficient C_P:

$$C_T = \frac{T}{1/2\rho U^2 \pi R^2}, \quad C_P = \frac{\Omega Q}{1/2\rho U^3 \pi R^2}, \tag{11.7}$$

where T is the turbine thrust, and Q the turbine torque.

11.3.2 Comparison Between PROPAN and ReFRESCO

This section presents the comparison of the calculations for the MCT at TSR = 5.15. An iterative pressure Kutta condition was applied in the panel method computations which, converged after three iterations to a precision of 10^{-3} for the non-dimensional pressure difference at the blade trailing edge. In the iterative convergence of the ReFRESCO computations, an L_2 norm of the normalised residuals lower than 10^{-5} is achieved for all quantities. For these residuals, convergence of the turbine forces is also obtained. We note that the computations take from minutes to hours with the panel method, in comparison with the RANS calculations that take from hours to days depending on the computer.

First, the flow streamlines obtained with code PROPAN are compared with the limiting streamlines obtained with code ReFRESCO in Fig. 11.3. An extensive region of separated flow is seen when approaching the trailing edge from the root to mid-blade in the RANS computations on the suction side of the turbine blade. Naturally, a fully attached flow is obtained with the panel method, which is not able to model flow separation. On the pressure side of the turbine blade, an almost two-dimensional flow is achieved with both methods.

Figure 11.4 illustrates the comparison of the pressure distributions on the turbine blade. The blade pressure is presented at the radial section $r/R = 0.30$, 0.50, 0.75, 0.90, 0.95 and 0.99. The pressure distributions are shown as function of the non-dimensional chordwise position s/c where s is defined along the cylindrical section chord for the turbine blade. A good agreement is seen for the turbine blade pressure distributions between the PROPAN and ReFRESCO computations, with some exceptions: different pressure distributions at sections $r/R = 0.3$ and $r = 0.5$ due to the occurrence of flow separation in the ReFRESCO computations, pressure oscillations near the leading edge which are linked to the transition prediction of the turbulence model, and in the blade trailing edge pressure distribution due to the differences in the geometries, where a sharp trailing edge geometry is assumed in the panel method computations in contrast to the ReFRESCO computations, where a round trailing edge geometry is considered. Details of the geometrical definition used in the present ReFRESCO calculations can be found in [17].

[1]Actually, a negative thrust force or drag.

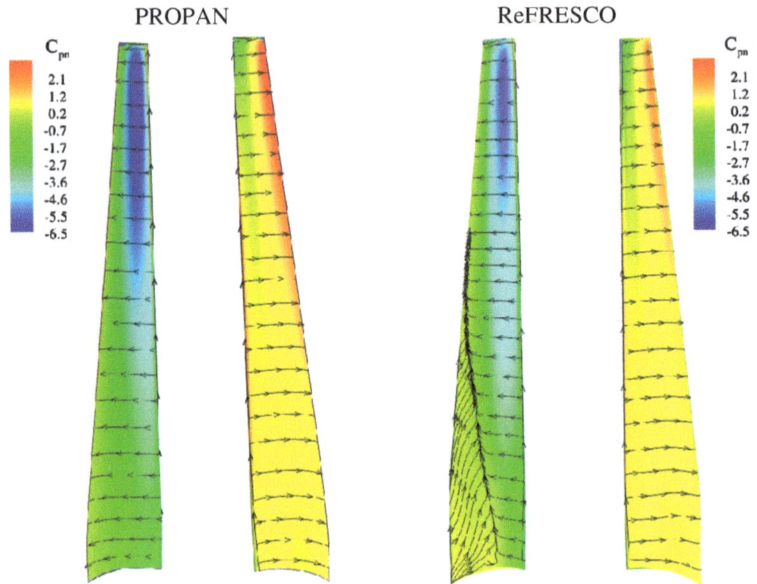

Fig. 11.3 Limiting streamlines on the pressure and suction sides of the turbine blade. Comparison between PROPAN and ReFRESCO for TSR = 5.15

To assess the wake alignment method in PROPAN, a comparison between the inviscid wake geometry and the viscous vorticity field at the planes $x/R = 0.1$ and 0.2 downstream from the turbine is presented in Fig. 11.5. A reasonable agreement is seen between the inviscid tip vortex and the region of high vorticity. However, identification of the viscous turbine wake is difficult to achieve due to the low grid resolution downstream of the turbine.

11.3.3 Comparison with Experimental Data

The comparison between the numerical results and the experimental data available from model tests is presented in Fig. 11.6. The experiments were performed in a towing tank at Southampton Institute for a carriage speed of 1.40 m/s and in the cavitation tunnel at QinetiQ, Haslar, Gosport, for a tunnel speed of 1.54 m/s. The experimental results have been corrected for blockage effects [2]. However, some differences between the towing tank and cavitation tunnel experimental results are observed.

Larger turbine thrust and torque are obtained with the panel code PROPAN in comparison with the viscous results and experimental data for the entire diagram. From these results two hypotheses are suggested to explain the differences. First, an inaccurate alignment of the blade wake with the potential flow model which

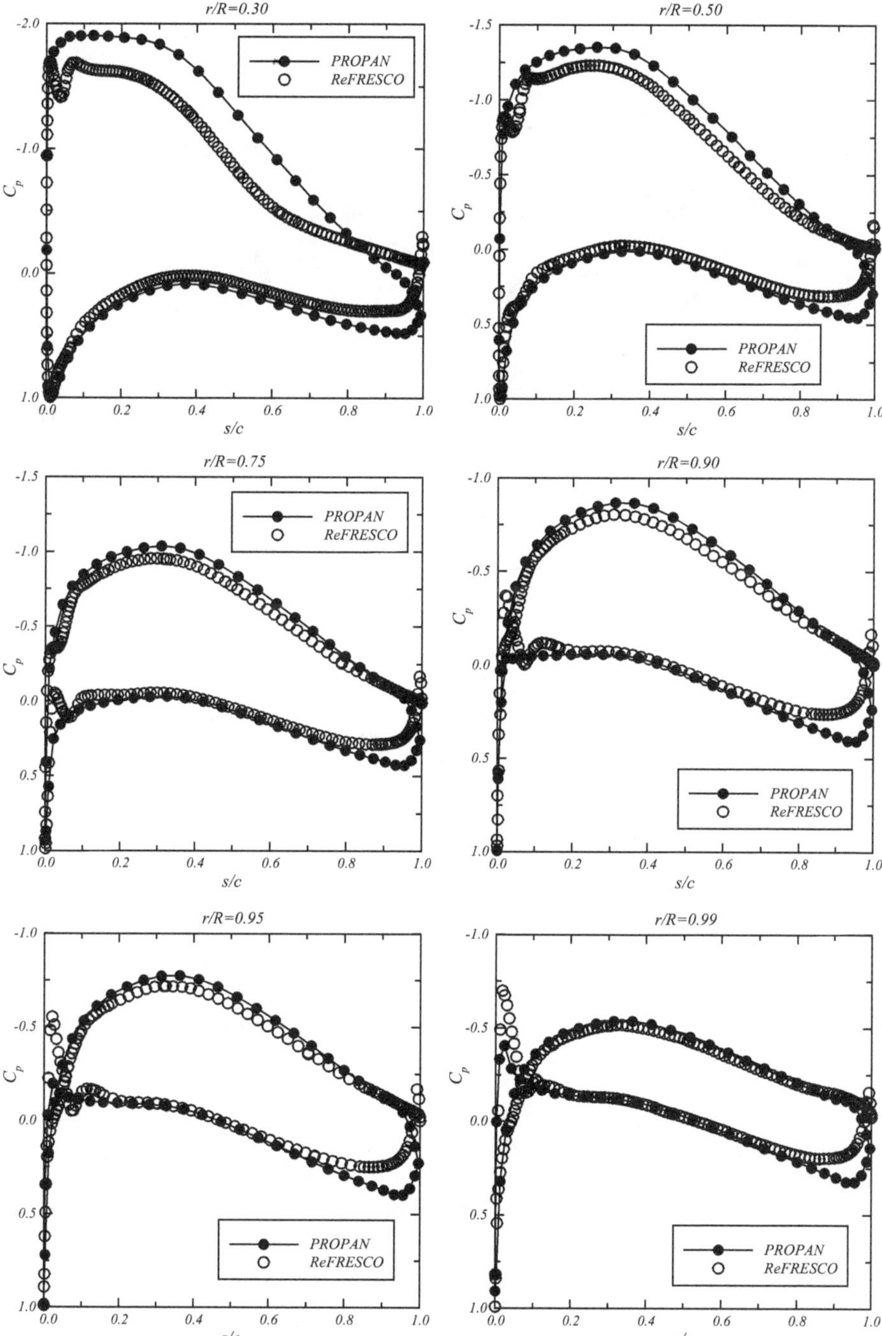

Fig. 11.4 Blade pressure distribution at the radial sections r/R = 0.30, 0.50, 0.75, 0.90, 0.95 and 0.99. Comparison between PROPAN and ReFRESCO for TSR = 5.15

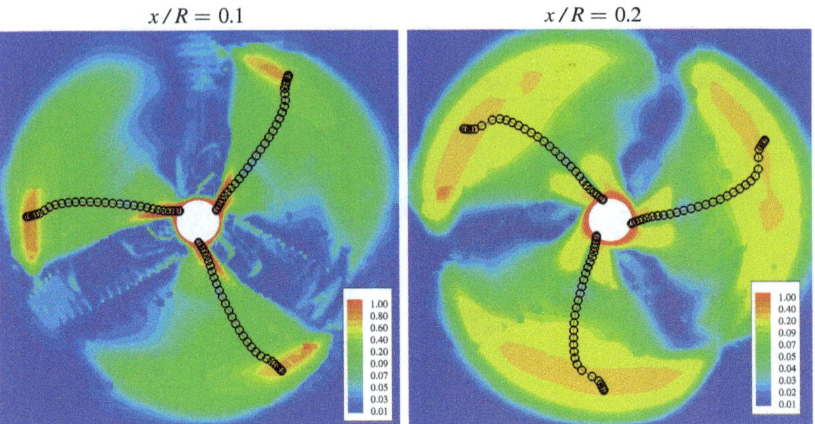

Fig. 11.5 Wake geometry at $x/R = 0.1$ (*left*) and $x/R = 0.2$ (*right*). Comparison between PROPAN and ReFRESCO for TSR = 5.15. The contours represent the ReFRESCO non-dimensional total vorticity field $|\boldsymbol{\omega}|/\Omega$ and the *symbols* represent the PROPAN wake geometry

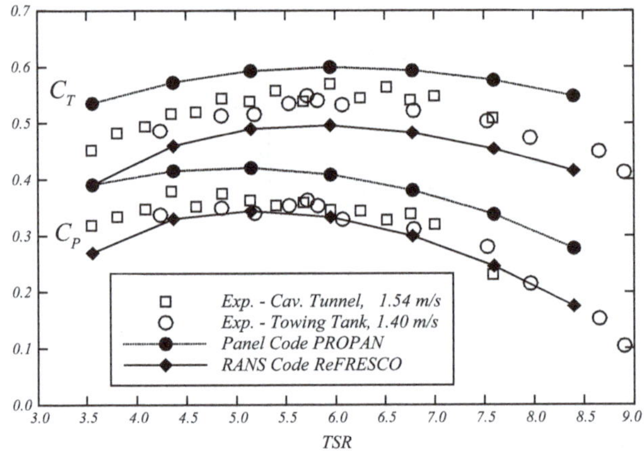

Fig. 11.6 Comparison between numerical and experimental results for a range of TSRs

influences considerably the turbine loading. The effect of the wake pitch on the loading of the turbine has been studied in [6]. Second, an under-prediction of the viscous corrections to the inviscid case. We note that viscous effects on the blade lift tend to decrease both the thrust and the power, and the effect of blade drag tends to increase the thrust and decrease the power.

A better agreement of the thrust and power coefficients is achieved with the RANS simulations in the comparison with the inviscid computations. Still, an under-estimation of the thrust and power is seen in the comparison with the experimental data. In the present calculations, transition is predicted near the blade leading edge by the turbulent model. However, no conclusion can be drawn about the predicted transition, since no information is provided by Bahaj et al. [2] for the flow regimes of the experiments. The RANS predictions tend to deviate from the experimental data for TSR = 3.56. For this TSR, a large separated region on the suction side of the turbine blade is obtained (shown by Otto et al. [17]), following what is already seen for TSR = 5.15 (Fig. 11.3). We note that the angle of attack to the blade sections increases with the reduction of the TSR. In this way, the use of a RANS solver to model the flow for TSR = 3.56 may be inadequate and the use of a URANS (Unsteady RANS) solver may be necessary to better predict the turbine forces, since it is able to model unsteady effects on the separated flow, that are likely to occur for this situation.

11.4 Conclusions

In this paper a comparison of panel method with RANS for uniform inflow conditions of a horizontal axis MCT using panel code PROPAN and RANS code ReFRESCO is made. The limiting streamlines, blade pressure distributions, wake geometry and thrust and torque coefficients between the two codes are compared. The turbine forces were also compared with the experimental data. The comparison shows the limitations of the inviscid flow model when the flow is separated. Still, a reasonable to good agreement of the blade distribution and turbine forces is seen between the models. The present study shows the advantages of the inviscid flow model for the preliminary design and analysis of an MCT, associated with the low computational time needed for routine design studies. When the flow is dominated by strong viscous effects, the use of a RANS solver becomes necessary for an accurate prediction of the flow, leading however to a substantial increase of the computational requirements.

Acknowledgements The authors acknowledge the use of MARIN code ReFRESCO in performing the current RANS computations. The authors would like to thank Prof. Luis Eça for his comments and suggestions, which helped to improve the manuscript. The authors also acknowledge the assistance of Mr. Joseph Cherroret in the numerical computations and post-processing analysis of the results.

References

1. Bahaj AS, Batten WMJ, McCann G (2007) Experimental verifications of numerical predictions for the hydrodynamic performance of horizontal axis marine current turbines. Renew Energ 32:2479–2490
2. Bahaj AS, Molland AF, Chaplin JR, Batten WMJ (2007) Power and thrust measurements of marine current turbines under various hydrodynamic flow conditions in a cavitation tunnel and a towing tank. Renew Energy 32:407–426
3. Baltazar J (2008) On the modelling of the potential flow about wings and marine propellers using a boundary element method. Ph.D. thesis, Instituto Superior Técnico

4. Baltazar J, Falcão de Campos JAC (2008) Hydrodynamic analysis of a horizontal axis marine current turbine with a boundary element method. In: Proceedings of the ASME 27th international conference on offshore mechanics and arctic engineering, Estoril
5. Baltazar J, Falcão de Campos JAC (2009) Unsteady analysis of a horizontal axis marine current turbine in yawed inflow conditions with a panel method. In: Proceedings of the first international symposium on marine propulsors, Trondheim, pp 186–194
6. Baltazar J, Falcão de Campos JAC (2011) Hydrodynamic analysis of a horizontal axis marine current turbine with a boundary element method. J Offshore Mech Arct Eng 133:041304-1–041304-10
7. Baltazar J, Falcão de Campos JAC (2012) Prediction of sheet cavitation on marine current turbines with a boundary element method. In: Proceedings of the ASME 31st international conference on ocean, offshore and arctic engineering, Rio de Janeiro 2012:1–11
8. Baltazar J, Falcão de Campos JAC, Bosschers J (2012) Open-water thrust and torque predictions of a ducted propeller system with a panel method. Int J Rotating Mach Article ID 474785 2012:1–11. http://www.hindawi.com/journals/ijrm/si/196407
9. Eça L, Hoekstra M (2012) Verification and validation for marine applications of CFD. In: Proceedings of 29th symposium on naval hydrodynamics (ONR), Gothenburg
10. HEXPRESS™ (2014) http://www.numeca.com/en/products/automeshtm/hexpresstm
11. Khan MJ, Bhuyan G, Iqbal MT, Quaicoe JE (2009) Hydrokinetic energy conversion systems and assessment of horizontal and vertical axis turbines for river and tidal applications: a technology status review. Appl Energy 86:1823–1835
12. Kinnas SA (1996) Theory and numerical methods for the hydrodynamic analysis of marine propulsors. In: Advances in marine hydrodynamics, computational mechanics publications. Southampton, Boston
13. Klaij CM, Vuik C (2013) Simple-type preconditioners for cell-centered, colocated finite volume discretization of incompressible Reynolds-averaged Navier-Stokes equations. Int J Numer Methods Fluids 71(7):830–849
14. Lawson MJ, Li Y, Sale DC (2011) Development and verification of a computational fluid dynamics model of a horizontal-axis tidal current turbine. In: Proceedings of the ASME 30th international conference on ocean, offshore and arctic engineering, Rotterdam
15. Menter F (1994) Two-equation eddy viscosity turbulence models for engineering applications. AIAA J 32(8):1598–1605
16. Morino L, Kuo C-C (1974) Subsonic potential aerodynamics for complex configurations: a general theory. AIAA J 12(2):191–197
17. Otto W, Rijpkema D, Vaz G (2012) Viscous-flow calculations on an axial marine current turbine. In: Proceedings of the ASME 31st international conference on ocean, offshore and arctic engineering, Rio de Janeiro
18. ReFRESCO (2014) http://www.marin.nl/web/Facilities-Tools/CFD/ReFRESCO.htm
19. Vaz G, Jaouen F, Hoekstra M (2009) Free-surface viscous flow computations. Validation of URANS code FRESCO. In: Proceedings of the ASME 28th international conference on ocean, offshore and arctic engineering, Honolulu, HI

Zeitfracht Medien GmbH
Ferdinand-Jühlke-Straße 7
99095 Erfurt, Deutschland
produktsicherheit@kolibri360.de